W. Kresse K. Fadaie
ISO Standards for Geographic

Springer
Berlin
Heidelberg
New York
Hong Kong
London
Milan
Paris
Tokyo

Wolfgang Kresse
Kian Fadaie

ISO Standards for Geographic Information

with 137 Figures and 13 Tables

 Springer

Dr. Wolfgang Kresse
FH Neubrandenburg
FB Bauingenieur-
und Vermessungswesen
Brodaer Str. 2
17033 Neubrandenburg
Germany

Dr. Kian Fadaie
Natural Resources Canada
588 Booth Street
K1A 0Y7 Ottawa
Canada

ISBN 978-3-642-05763-2

Springer-Verlag Berlin Heidelberg New York

Cataloging-in-Publication Data applied for

Bibliographic information published by Die Deutsche Bibliothek
Die Deutsche Bibliothek lists this publication in the Deutsche Nationalbibliografie;
detailed bibliographic data is available in the Internet at <http://dnb.ddb.de>.

This work is subject to copyright. All rights are reserved, whether the whole or part of the material is concerned, specifically the rights of translation, reprinting, reuse of illustrations, recitation, broadcasting, reproduction on microfilm or in any other way, and storage in data banks. Duplication of this publication or parts thereof is permitted only under the provisions of the German Copyright Law of September 9, 1965, in its current version, and permission for use must always be obtained from Springer-Verlag. Violations are liable for prosecution under the German Copyright Law.

Springer-Verlag Berlin Heidelberg New York
a member of BertelsmannSpringer Science+Business Media GmbH

http://www.springer.de

© Springer-Verlag Berlin Heidelberg 2010
Printed in Germany

The use of general descriptive names, registered names, trademarks, etc. in this publication does not imply, even in the absence of a specific statement, that such names are exempt from the relevant protective laws and regulations and therefore free for general use.

Product liability: The publishers cannot guarantee the accuracy of any information about the application of operative techniques and medications contained in this book. In every individual case the user must check such information by consulting the relevant literature.

Camera ready by authors
Cover design: E. Kirchner, Heidelberg
Printed on acid-free paper   30/3141/as 5 4 3 2 1 0

To our children

Randi and Gregor in Neubrandenburg

Payam in Ottawa

Acknowledgements

We as the authors say thank you to our colleagues and friends who have contributed their expertise, knowledge, and skills to this book and who have helped us to get it completed. We also thank our spouses, Almut and Bahram Khoee for being patient with us.

Astrid Balada, DIN, Berlin, Germany
Morten Borrebæk, Norwegian Mapping Authority, Oslo, Norway
Sybille Brozio, FH Neubrandenburg, Germany
Steve Carson, GSC Associates, Las Cruces, New Mexico, USA
Antony Cooper, CSIR, Pretoria, South Africa
Wyn Cudlip, Qinetiq, United Kingdom
David Danko, ESRI, Vienna, Virginia, USA
Ulrich Düren, Landesvermessungsamt Nordrhein-Westfalen, Bonn, Germany
Jürgen Ebbinghaus, AED-SICAD, München, Germany
Ernst Heil, FH Neubrandenburg, Germany
John Herring, Oracle, Reston, Virginia, USA
Yves Hudon, Gouvernement du Québec, Canada
Hiroshi Imai, University of Tokyo, Japan
Helen Kerfoot, Natural Resources Canada, Ottawa
Roger King, Mississippi State University, USA
Elfriede Knickmeyer, FH Neubrandenburg, Germany
Hans Knoop, DIN, Berlin, Germany
Andy Kwan, Industry Canada, Ottawa
Hans-Jürgen Larisch, FH Neubrandenburg, Germany
William McGrum, Industry Canada, Ottawa
David McKellar, Department of National Defence, Ottawa, Canada
Martin Nitschke, FH Neubrandenburg, Germany
Douglas O'Brien, IDON, Ottawa, Canada
Olaf Østensen, Norwegian Mapping Authority, Oslo, Norway
Bettina Petzold, FIG, Stadt Wuppertal, Germany
Clemens Portele, Interactive Instruments, Bonn, Germany
Erhard Pross, Bundesamt für Kartographie und Geodäsie, Leipzig, Germany
Dora-Inés Rey, Institut Agostin Codazzi, Bogotá, Colombia
Charles Roswell, National Imagery and Mapping Agency, Reston, Virginia, USA
John Rowley, Geobase, Epsom, Surrey, United Kingdom
Bjørnhild Sæterøy, Norwegian Technology Centre, Oslo, Norway
Paul Smits, Joint Research Centre, Ispra, Italy
Roger Stacey, Ottawa, Canada
Marguerite Trindade, Canada Centre for Remote Sensing, Ottawa
Herman Varma, Halifax, Canada

Contents

Introduction 1

1 Basics of standards 3
  1.1 Characteristics of standards 3
  1.2 International standardisation organizations and consortia 8
  1.3 Formal international standardisation organizations 9
      1.3.1 International Organization for Standardisation (ISO) 9
      1.3.2 International Electrotechnical Commission (IEC) 18
      1.3.3 ISO/IEC Joint Technical Committee 1 (JTC1) 18
      1.3.4 International Telecommunication Union (ITU) 18
      1.3.5 Access to standard-documents 19
  1.4 ISO 9000 family of standards 19
  1.5 Cultural and linguistic adaptability 22
  1.6 Requirements for future developments of standards 24

2. Geomatics standards 27
  2.1 History of geomatics standards 27
  2.2 ISO/TC211 30
      2.2.1 History and work of ISO/TC211 30
      2.2.2 Example of an ISO-compliant GIS 35
      2.2.3 Reference model (ISO 19101) 41
  2.3 Future of geomatics standardization 48
      2.3.1 Implementation strategies 48
      2.3.2 Ongoing work of ISO/TC211 49
      2.3.3 Directions and Activities 50
  2.4 Roadmap to the ISO 19100 standards 53

3. Non-geometry standards 55
  3.1 Infrastructure standards 55
      3.1.1 Conceptual schema language (ISO/TS 19103) 55
      3.1.2 Terminology (ISO 19104) 66
      3.1.3 Conformance and testing (ISO 19105) 67
      3.1.4 Profiles 68
  3.2 Basic standards 71
      3.2.1 Services (ISO 19119) 72
      3.2.2 Rules for application schemas (ISO 19109) 75
      3.2.3 Spatial schema (ISO 19107) 76
      3.2.4 Temporal schema (ISO 19108) 76

3.2.5 Georeference 77
  3.2.6 Portrayal (ISO 19117) 85
  3.2.7 Encoding (ISO 19118) 87
  3.2.8 Metadata (ISO 19115) 89
  3.2.9 Quality (ISO 19113, ISO 19114, ISO/TS 19138) 93
 3.3 Imagery standards 96
  3.3.1 Framework (ISO 19129, ISO/TR 19121, ISO/RS 19124) 96
  3.3.2 Sensors (ISO 19130) 102
  3.3.3 Sensor Model Language (SensorML) 108
  3.3.4 Coverages (ISO 19123) 110
 3.4 Catalogue standards 111
  3.4.1 Procedure for registration (ISO 19135) 111
  3.4.2 Feature catalogues (ISO 19110) 112
  3.4.3 Data dictionary registers (ISO 19126) 114
 3.5 Implementation standards 115
  3.5.1 Data product specification (ISO 19131)............................... 115
  3.5.2 Simple features (ISO 19125) 115
  3.5.3 Web Map server interface (ISO 19128) 115
  3.5.4 Location based services (ISO 19132 – ISO 19134) 117
 3.6 Qualifications and certification of personnel (ISO/TR 19122) 123

4. Geometry standards 125
 4.1 Relations between the geometry standards 125
 4.2 Positions 126
 4.3 Spatial schema (ISO 19107) 128
  4.3.1 Overview over ISO 19107... 128
  4.3.2 General description of the geometry classes of ISO 19107 129
  4.3.3 Detailed description of the geometry classes of ISO 19107 132
 4.4 Simple Features (ISO 19125-1) 159
  4.4.1 Point and MultiPoint 160
  4.4.2 Curve and MultiCurve 160
  4.4.3 Polygon and MultiPolygon 162
  4.4.4 Example for the topology in ISO 19107 and ISO 19125-1 163
 4.5 Schema for coverage geometry and functions (ISO 19123) 165
  4.5.1 Overview over ISO 19123 166
  4.5.2 Description of the discrete coverage classes of ISO 19123 166
  4.5.3 Description of the continuous coverage classes of ISO 19123 169
  4.5.4 Operations in ISO 19123 174
 4.6 Geography Markup Language (GML) (ISO 19136) 177
  4.6.1 GML schemas 179
  4.6.2 GML application schema 183
  4.6.3 Relation between GML, OGC, and the ISO 19100 standards 183
  4.6.4 Alignment between ISO 19136 and GML-development 184

5. Liaison members of ISO/TC211 185
 5.1 Internal liaison members of ISO/TC211 185
  5.1.1 ISO 187

5.1.2 IEC	189
5.1.3 ISO/IEC JTC1	190
5.1.4 ITU	192
5.2 External liaison organizations to ISO/TC211	192
5.2.1 Portrait of all (status July 2003)	192
5.2.2 Open GIS Consortium (OGC)	207

6. Applications 213
  6.1 Canadian GIS industry 213
  6.2 German standard-based systems for cadastral and topographic
      information 214

Bibliography 217
Annex A: Terms and definitions of the ISO 19100 standards 225
Annex B: ISO 19115 Metadata package data dictionaries 260
Annex C: Extensible Markup Language (XML) 286
Annex D: Abbreviations 299
Annex E: Class names 307
Annex F: Past and planned meetings 308
Index 309

Introduction

The worldwide official standards are developed by the International Organization for Standardisation (ISO) and two other similar organizations. The ISO/Technical Committee 211 (ISO/TC211) has developed the ISO 19100 family of standards, named Geographic information / Geomatics. This book is a tutorial for the ISO 19100 standards explaining their meaning, interrelation, and origins. Some national examples illustrate their application.

The book is addressed to technical experts who already have some background knowledge in Geographic Information Systems (GIS) and who want to know more about standardisation in GIS, in particular, the role of the ISO. The book also addresses the needs of programmers who are going to implement ISO 19100 standards and need a better understanding of the overall structure of the standards. Last, but not least, the book is intended for teaching the basics of GIS. Many of the illustrations will help to better understand some of the rather abstract ISO documents.

Chapter 1, sections 1.1 – 1.3 introduce the standardisation language and explain the role of a worldwide "Standards Developing Organization". This is important for understanding the original ISO documents, particularly if one intends to take part in the standardisation process. Section 1.4 is dedicated to the well-known ISO 9000 family of standards, section 1.5 addresses the linguistic adaptability of the standards, and section 1.6 provides an outlook on future requirements for standards.

Chapter 2, section 2.1 reviews standards that have gained some relevance for geomatics in the past, including the roots of ISO/TC211 in Europe. Section 2.2 gives an overview over the work of ISO/TC211 including an example-GIS and the description of the reference model. Section 2.3 discusses possible future directions of GIS standardisation, and changes in the relevance of GIS as a whole.

Chapter 3 is one of the two main bodies of the book. The chapter starts with an explanation of the Information Technology (IT) background to the geomatics standards and is followed by a description of all non-geometry standards. If the reader requires more details than those presented here or even the full text of the document, ISO or other sources have to be contacted.

Chapter 4 is the other of the two main bodies of the book and explains the details of the geometry-oriented ISO 19100 standards. The chapter is intended primarily for GIS instructors.

Chapter 5 introduces the partner organizations to ISO/TC211 that are referred to as liaison members. One of the most important liaison members is the Open GIS Consortium (OGC). This chapter illustrates the central role of ISO/TC211.

Chapter 6 contains a few examples that demonstrate the wide range of possible applications based on the ISO 19100 geomatic standards.

1 Basics of standards

1.1 Characteristics of standards

Although everybody recognises that standards have become essential in every corner of our life, they are usually considered a dry-as-dust topic that seems to block inspiration and flexibility. Standards are assumed to be a weak compromise between existing and proven solutions developed by dull administrative people.

In reality, standardisation is a real challenge often comparable to the most sophisticated development projects in the industry. Good standards are simple and unique. For example, Roman letters are in use after more than 2000 years and they are still able to adapt to almost any language of the world. In contrast, the Roman numbers were not that successful as a standard because the theory behind was not mature enough. As everybody knows, a numbering system without a zero has limited applications.

All of us have gained some experience with standards during our lifetime. An example of a standard is A3 and A4 paper sizes used in most parts of the world, or the letter and legal paper size standards used in America. This kind of a standard is helpful and today, nobody would start arguing about the standard paper sizes.

Another group of examples are for file formats. The Microsoft Word *.doc format is well known. Images are often stored or transferred in TIFF. We use them regularly, or rather we let our computer use them. These formats have been developed in conjunction with particular computer programs. The *doc format belongs to Microsoft's Windows word processing system. The TIFF was created by Aldus and adopted by Adobe.

A further example of a standard is the ISO 9000 series for quality management. These define a set of rules for investigating the efficiency of quality management in an organisation. The ISO 9000 standards are more abstract than the other examples but they are enormously important economically for the companies being tested as well as for the accredited companies responsible for the certification.

These examples illustrate the vast and heterogeneous nature of standardisation. Standards can be technical or management-oriented, detailed or abstract, the result of an international consensus-building process or of one single company's development.

In order to move closer to the world of standardisation, the subject of our interest will be viewed from different perspectives.

Linguistic perspective and types of standards

In the Medieval Europe, a standard was a flag or a sculptured object with the distinctive ensign of a king raised on a pole to indicate the rallying point of an army. It was also later used for the authorized example of a unit of measure or weight (OED 1970).

- In modern English the word "standard" has a number of different meanings, even within the standardisation business: An official standardisation organization like the ISO publishes formal standards that have a certain level of relevance in the application domain. Often this type is called a *de-jure* standard. In the French and German languages the corresponding terms are "la norme" and "die Norm".
- Companies develop industry standards for the purpose of operating their products. These standards are not officially branded unless they put them through the official standardisation process. If the standard is well accepted by the user community it becomes a so-called *de-facto* standard.
- In sectors like the IT-business and geomatics, consortia of companies agree on common specifications to ensure interoperability. These specifications are also *de-facto* standards. However, in recognition of the companies' intention to develop common technical rules, they are called standards from the beginning.

Economic perspective

The drive for industry to invest in standardisation efforts is to benefit from the tremendous savings of money that ensue. A study in the year 2000 proved that the economic advantage of standards for the German industry is 15 Billion Euros annually (Hartlieb 2000). The reasons in detail are:

- Standardisation avoids the costs of adapting interfaces to a range of applications.
- Participation in the standardisation process puts companies ahead of others not participating in the process.
- Standards enable a company to utilise a range of suppliers rather than become reliant on a limited number of sources.
- Standards support the legislation process. About 20% of the German standards (DIN) are referenced by laws and by-laws and take the burden of solving detailed technical questions from the legislation body. In this sense the standardisation simplifies the legislation process as the parliament does not have to deal with the subject in full details. The abstract laws only refer to the DIN-standards that cover the details.

User's perspective

If a technology is mature, then users expect standardised solutions that are simple, fast, and effective. For example, in 1996 a production company looking for a data exchange format that suited all the practical needs for airborne imagery type data initiated the ISPRS Working Group dedicated to standardisation. This user-driven case is a typical starting point for standardisation.

After the initiative had started the discussions in the ISPRS WG and in other committees discovered that a standard had to be designed in a much broader sense than originally intended.

A well-known format for referencing imagery to the earth is GeoTIFF. It is a good solution for small scale or geo-rectified imagery but has some strong limitations. GeoTIFF is based on a well-defined set of TIFF-tags. The original idea was to simply extend the set of TIFF-tags to meet the requirements of airborne photogrammetry. However, it turned out that a simple definition of additional TIFF-tags would only foster a temporary solution as new sensor types continuously show up on the market. As programming time is expensive one would not decide for interim solution. A desirable extension of the subject towards generic transformations and general sensor geometries is not possible without having developed an extended theoretical foundation and strategies to adapt the implementation to the latest solutions with a minimum effort.

In fact the modern solutions are model driven based on the ISO 19100 standards. The implementation will be using the Extensible Markup Language (XML). One of the broader packages is the Sensor Model Language (SensorML) discussed in this book.

System manufacturer's perspective

The best solution for a system manufacturer is a closed universe based on the manufacturer's system and without any interfaces to the outside world. Today, this is no longer realistic. Government agencies and other customers have forced the manufacturers to open their systems and support standardised interfaces. In a continuously expanding market such as geomatics, it seems that sharing resources with industrial partners can favour the economic growth of a company.

If standardisation is pushed towards an implemented solution the companies not only have to pay for drafting the standard's documentation but also for programming the implementation. This can be expensive. In order to guarantee a return of investment, companies in the IT-business (including geomatics) have opted for joint solutions. The best-known example is the Open GIS Consortium.

Development group's perspective

A development group, usually called the standardisation committee, has to draft a document that contains a solution for the standardisation task. The document would also address future challenges, the details of which are often unknown when the development takes place. In this sense standardisation, particularly in the IT-domain, is very closely related to the design of computer systems. A lot of money is invested according to the content of a specification and the better the specification, the longer the investment will hold.

Choosing the right moment to launch a standard development is probably the most critical decision in the process. If it starts too early, newer industrial developments will overtake the standardisation documents. If it starts too late, there might be little room left for the standardisation among existing solutions that have established a dominant position.

As standardisation generally takes place *after* industry has completed a number of developments, the standardisation committee is always confronted with existing solutions mostly with limited compatibility. Most software companies are not willing to invest in software adaptation because the modification of existing software towards compliance with a new standard is extremely expensive. This has accelerated the creation of industrial consortia in the IT-domain where common industry standards are often defined before the investment in the software development starts. The development of a sufficiently generic standard in combination with the ability to define profiles for specific fields of application is a widely used strategy to integrate existing software into new standards. This strategy leads to compatibility on the abstract level but not necessarily on the implementation level. Often, an extended compatibility can only be reached with future versions of the application software or with the advent of newer technologies. An example is the Extensible Markup Language as the *de-facto* standard of today for the encoding of documents and database contents.

National perspective

The principles of standardisation are: do it once, do it right, do it internationally. In the IT-domain any national approach is at best a preparation for or a benchmarking of an international solution. A typical example of this process is the history of the European approach to the standardisation of geomatics, the CEN/TC287 (Geographic Information) (CEN = Comité Européen de Normalisation). This committee was dissolved after the ISO/TC211 started its work and it's members expertise, and knowledge, were integrated into the new worldwide committee.

However, International Standards do not mean that national or regional distinctions are overridden. An important subject is the linguistic adaptability of the standards. As the English language dominates business in the world, in particular the IT sector, it is often difficult to keep the coherence of a standard-compliant system and a national user interface. Admittedly, it is often a matter of costs. For further details see chapter 1, section 5.

Differences in traditions are more difficult to be solved. The first North American geographic information systems were designed for medium scale applications. From the beginning, they permitted integration of remote sensing technology. The European systems followed their cadastral tradition that is more than two centuries old. America thinks in feet. Europe thinks in centimetres. This causes different viewpoints during the standardisation development.

Hierarchy of standards

Every standard has its typical level of detail that must be agreed upon before the work starts. Usually the IT-domain recognises three levels: abstract, implementation, and interface. Standards at the abstract level are independent of operating systems, applications, hardware, and encodings. Standards at the implementation level determine encoding. Standards at the interface level determine hardware or hardware-oriented software, often called firmware. As a general principle, ISO standards reside at the abstract level. Most industry standards are at the implementation level or the interface level.

Classification perspective

The following section demonstrates an alternative approach to open up the subject of standardisation. This approach defines a grouping of standards that covers far more than the formal standards addressed by the ISO. In fact, the intent is to be able to include any technical rules applicable in industry and administration. This grouping is hierarchically organized in two levels.

Table 1.1. Classification of standards (Kim et al. 2002)

Top level	Bottom level	Explanation
Level of coverage	International Multinational, Regional National Local	
Level of prescriptiveness	Recommended practice Information report Standard	leads to an advisory document informative document normative document
Function	Design standards	focus on user consistency in respect to product structure and appearance. Examples for design standards are zip codes and metadata. They still leave reasonable room for innovation
	Interface standards	facilitate the interconnection of components or adjoining systems like communication protocols and plug compatibilities

Table 1.1. (cont.)

	Framework standards	are the foundations of multiple products and services like the metric system, data dictionaries, and a Coordinate Reference System
	Performance standards	are results oriented and do not specify how to do it. Examples are the braking distance or the pavement service life
	Testing methods	provide consistent and replicable methods for assuring quality and compliances. Examples are crash tests, conformance testing, and ISO 9000
	Terminology	creates agreements on what words mean, accelerating contracting and minimizing confusion and conflict
Development process	De-facto standards	arise from market forces. The de-facto standards are most successful when dominant participants can "dictate" standards. But they also help speed the entry of smaller competitors, especially when standards are open and not proprietary
	Regulatory standards	are created and enforced by public agencies through rule-making. These standards are best for public safety and health or for situations where economics require system-wide actions like the emission standards for automobiles
	Consensus standards	are either voluntary agreements or set via Standards Developing Organizations (SDOs) like the ISO

1.2 International standardisation organizations and consortia

An individual trying to find the standard that is relevant to an application may encounter a large number of groups and organizations that claim to be competent in the standardisation business. At the international level ISO and OGC (Open GIS Consortium) are well known to the geographic information community. However, other acronyms like IEC and ITU are less familiar. In addition, the standards community works with a lot of abbreviations, like TC, SC, JTC, WG, SIG that makes them difficult to read and understand. Nevertheless, once the meaning of the abbreviations is known they can simplify communication. This section reviews the important groups, explains their terminology, and describes their interrelations.

The groups can be subdivided into two categories, *international organizations* and *international consortia*.

International organizations base their decisions on consensus. By having a budget scheme that allocates the financial burden of the organization to all member countries according to their economic potential, international organizations are fairly independent of the interests of individual governments or industries. Most of the organizations have a long history - some of them are almost 150 years old - today their work-rate is sometimes considered to be too slow for industry needs.

Three international organizations dominate the field of standardisation. They are sometimes called the Standards Developing Organizations (SDO): the ISO is the "International Organization for Standardisation", the IEC is the "International Electrotechnical Commission", and the ITU is the "International Telecommunication Union".

The members of *international consortia* are drawn primarily from industry, often from government agencies, and universities are occasionally represented. The primary goal of international consortia is to bundle the interests of their members. One of their interests is the development of common standards in order to advance other developments. Though the standard development is done with the participation of all members, the strong influence of the larger companies cannot be neglected. Though the standards might not be built on a broad consensus, the results are generally technically feasible and foster progress on the subject. They are called industry or *de-facto* standards.

The Open GIS Consortium (OGC) could presently be considered the most important consortium in the geographic information community. Another example of a consortium is the W3C, the "World Wide Web consortium". It organizes the work necessary for the development and evolution of a Web technology into activities.

1.3 Formal international standardisation organizations

Originally the three organizations ISO, IEC, and ITU had their well-defined fields of activity. The IEC dealt with the electro technical equipment. The ITU dealt with the radio transmission, and the ISO was responsible for all other subjects. However, today's sectors such as computer science have developed and the borders between the subjects have become blurred. There are current demands to merge the work of all three organizations into one enterprise.

1.3.1 International Organization for Standardisation (ISO)

The "International Federation of the National Standardising Associations" (ISA) was founded in 1926 with its primary focus on mechanical engineering. The ISA ceased its activities in 1942, owing to World War II. In 1947 the ISO was established as a new non-governmental organization and continued the work. The ISO Central Secretariat is in Geneva, Switzerland.

People are sometimes confused by the mismatch between the name "International Organization for Standardisation" and the three letters ISO. In fact, the name of ISO is not an acronym, but rather a word derived from the Greek *isos*, meaning "equal", which points to one of the goals of international standardisation. This name is used around the world to denote the organization thus avoiding the plethora of acronyms resulting from the translation of the full name into many different languages.

The work of ISO is based on three principles:

1. Consensus

The views of all interested parties are taken into account. This is sometimes referred to as the democracy within the standardisation development. In practice, the influence in shaping a technical standard is largely restricted to the parties that can afford to pay for their experts to be involved. However, the final vote on a draft for an International Standard depends on an agreement of 75% of the voting parties, independent of their activities during the development process.

2. Industry-wide

The standards will always lead to global solutions that satisfy the needs of industries and customers worldwide. Due to diverse developments in different parts of the world this principle often leads to a minimum consensus and to abstract-level standards. If worldwide consensus on a specific subject turns out to be impossible, the ISO would then withdraw its involvement.

3. Voluntary

International standardisation is market-driven and therefore based on the voluntary involvement of all interested parties. If a technical subject requires standardisation, the interested parties would approach ISO asking for guidance on the standardisation process. ISO will guarantee compliance with the consensus principle and consistency with other International Standards. The development will then receive recognition as an international ISO standard.

1.3.1.1 Members of ISO

ISO is the umbrella organization for national standardisation activities. ISO members are primarily countries which are usually represented by their respective national standards organizations. Examples of such national bodies are the American National Standards Institute (ANSI), the Standardisation Council of Canada (SCC), and the DIN (Deutsches Institut für Normung).

The national bodies are often unable to provide the complete range of expertise required for standard setting. Therefore, scientists and engineers usually exchange their knowledge within dedicated international organizations like the International Cartographic Association (ICA) or the International Society for Photogrammetry and

1.3 Formal international standardisation organizations 11

Remote Sensing (ISPRS). Companies express their interests through consortia like the Open GIS Consortium (OGC). In order to incorporate this expertise, ISO has created another membership type called the external liaison organization.

In an effort to maintain consistent standards across committee borders and to avoid duplication of work, formal relations are established among ISO Technical Committees or Subcommittees dealing with similar subjects. Theses relationships are called internal liaisons.

1.3.1.2 Technical Committees (TCs) and Subcommittees (SCs)

The work of ISO is decentralised. While the ISO Central Secretariat is mainly concerned with the development of consistent standards according to the ISO regulations, the actual work is done in the Technical Committees (TC) and the Subcommittees (SC). Every TC and SC has its own secretariat at a competent institution somewhere in the world. For example, the secretariat for the ISO/TC211 (Geographic information / Geomatics) is attached to the Norwegian Mapping Authority near Oslo. The chairman has a powerful position within the TC because he decides on the direction of the TC.

The number of members in a TC may range from 10 to 1,000. In the case of TC211, there are presently about 200 members that meet about twice a year. In addition, today's communication technology allows almost daily exchange of notices and documents.

Bjørnhild Sæterøy is project manager at the Norwegian Technology Centre near Oslo. On the international level she is the secretary of the ISO/Technical Committee 211 (Geographic information /Geomatics). In her position she keeps this group of 200 experts moving, acknowledged and being liked by probably everybody in the committee. The educational background of Mrs. Sæterøy is business and technical. She studied accounting, business administration and data programming. Her experiences range from research and software development to the work in an idealistic organization. She worked in a publishing house and for maritime companies.

Mrs. Sæterøy's first interaction with standardisation happened in 1992 when she became the Norwegian representative to the ISO/IEC JTC1 (Information Technology). She is presently secretary of several national committees related to ISO and ISO/IEC JTC1 apart from her position as secretary of ISO/TC211.

When we asked her why she is being engaged in the ISO/TC211 works, she briefly said: Who could resist?

A Technical Committee is subdivided into Working Groups (WGs) that are usually responsible for the development of an individual standard or similar deliverable of the ISO. If the subject is very broad, the WG may be subdivided into two or more project teams that then become responsible for a single standard. For example, the standard ISO 19115 (Metadata) has been developed by a project team within the ISO/TC211 WG 3 (Geospatial Data Administration).

If a TC becomes very large it has been found to be advisable to put the work of chairing the committee on more than one pair of shoulders. It is for this reason the ISO has established Subcommittees (SCs). A TC may have many, one, or no SCs. It is a matter of power and politics whether a SC is established or not, because the chairman has essentially the same influence as the chairman of a TC.

The TCs and the SCs are standing committees which enables them to provide the maintenance that a standard requires in a long-term perspective. In contrast, the project teams and Working Groups are dissolved once a standardisation project has matured. The TC and the SC initiate a review of each standard after at most, a five year interval. The outcome of this review may be confirmation, revision, or withdrawal of the standard.

The long-term cooperation of the TC-members fosters personal contacts and often leads to new technical ideas. A TC is always open to new members and expertise and can be compared to a large family. At present, ISO has 186 Technical Committees and less than 50 Subcommittees.

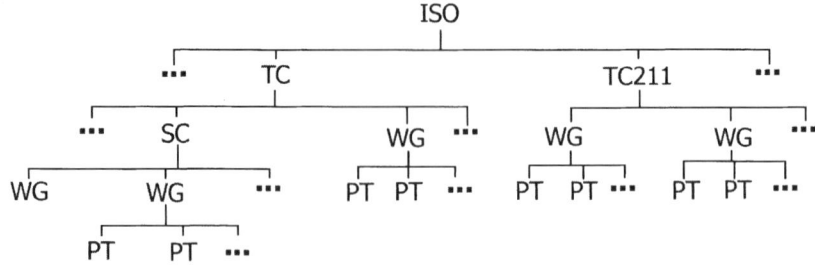

Fig. 1.1. Structure of ISO-committees: TC, SC, WG, Project Team (PT)

The ISO/TC211 has no Subcommittees (SC). Other Technical Committees such as ISO/TC20 (Aircraft and space vehicles) have some Subcommittees that function as another structural level.

1.3.1.3 Standardisation procedure in ISO

The idea for a new standard is always born outside ISO and the ISO's role is to formalise a method to forward the idea to an internationally approved standard. The steward of the work is a Technical Committee or a Subcommittee. The development process for a standard is subdivided into six consecutive stages.

1. **Proposal stage**. If an industry sector identifies the need for a new standard then the industry representatives will ask its national standardisation organization to send a New Work Item Proposal (NWIP) to ISO. The ISO Central Secretariat assigns this New Work Item Proposal to the most relevant Technical Committee (TC) or Subcommittee (SC) and sends it out for a vote. It is only if the majority of the P-members of the TC or the SC vote in favour of the New Work Item Proposal and at least five national bodies agree to actively participate in the Working Group, that ISO will advance the process. During this stage the project team is formed and the members reach a consensus about the objectives of the planned standard.

2. **Preparatory stage**. The TC or SC sets up a Working Group and identifies a chairman referred to as the convener. The Working Group is asked to prepare a working draft (WD) as a first written summary of the future standard. Successive working drafts may be considered until the Working Group is satisfied that it has developed the best technical solution to the problem being addressed. At this point, the draft is forwarded to the Working Group's parent committee for the consensus-building phase. During this stage the first version of the standard is developed.

3. **Committee stage**. As soon as a first committee draft (CD) is available, the ISO Central Secretariat registers the document. It is distributed to all members of the TC or SC for comments. Within this stage the document is reviewed carefully by independent experts that have not taken part in the internal discussions. During this stage the standard reaches maturity.

4. **Enquiry stage**. In this stage the document becomes a Draft International Standard (DIS). Over a period of five months, the ISO Central Secretariat circulates the document to all ISO member bodies for voting and comments. It is only approved for submission as a Final Draft International Standard (FDIS) if a two-thirds majority of the P-members of the TC or SC are in favour, and not more than one-quarter of the total number of votes cast are negative. If the approval criteria are not met, the text is returned to the originating TC or SC for further study. During this stage the standard is aligned with the existing standards of ISO and with other ongoing work.

5. **Approval stage**. The ISO Central Secretariat circulates the Final Draft International Standard (FDIS) to all ISO member bodies and requests that a final Yes/No vote be made within a period of two months. Technical comments that are received during this period are not considered, but will be registered for consideration during future revisions of the International Standard. The text is approved as an International Standard if a two-thirds majority of the P-members of the TC or SC are in favour and not more than one-quarter of the total number of the votes cast are negative. If these approval criteria are not achieved, the standard is referred back to the originating TC or SC for consideration in the light of the tech-

nical reasons submitted in support of the negative votes cast. During this stage the standard has fulfilled all formal ISO requirements.

In order to meet requirements for an accelaration of the formal standidisation procedure the ISO Central Secretariat has started to weaken the rules and eventually publishes a positively voted DIS as an International Standard, thus skipping the approval stage if major comments are not anticipated.

6. **Publication stage**. Once a Final Draft International Standard has been approved, only minor editorial changes are made to the final text. The ISO Central Secretariat publishes the International Standard (IS).

When difficulties are encountered in achieving a consensus, the standardisation process may take up to five years. Many standards are completed just prior to this deadline in order to avoid all the efforts of the team being officially deleted. If the parties still require the standard after the five year deadline has passed, then the whole process has to start over again.

1.3.1.4 Deliverables of ISO

The ISO is commonly considered an organization that makes standards or more precisely, leads their development and finally publishes them as binding International Standards (IS). However, this is only partially true. Firstly, until around 1970, ISO only published international recommendations. Secondly, ISO has defined a reservoir of tools with simpler development procedures than full International Standards. These other tools may be taken into consideration unless a full International Standard is out of scope as a result of slowness, lacking consensus, or simply a different intention like informative reporting.

The products of ISO are standard-documents. They are generally called deliverables. The following table shows a complete list of ISO-deliverables:

- ISO Standard
- ISO/PAS – ISO Publicly Available Specification
- ISO/TS – ISO Technical Specification
- ISO/TR – ISO Technical Report
- IWA – International Workshop Agreement

ISO Standard

An ISO Standard is normative. It is assumed that the stages one through three of the standard development process are completed. In stage four, the enquiry stage, the ISO Central Secretariat sends out the Draft International Standard (DIS) to all ISO-members for vote within a five month period. If two thirds of the voting members (P-members) vote in favour and no more than one quarter vote against, the DIS is approved. It moves on to stage five, the approval stage. After integrating comments, the document is issued as a Final Draft International Standard (FDIS) for a final vote with the same voting procedure as with the DIS, but with a period of only two months. If approved, it is published as an International Standard (IS) with a review of the IS taking place at least every five years.

ISO/PAS – ISO Publicly Available Specification

An ISO Publicly Available Specification (ISO/PAS) is also normative. After completion of stage two of the standard development, the preparatory stage, the Working Group must find consensus on the document. A simple majority of the P-members of the Technical Committee (TC) or Subcommittee (SC) is then necessary to create a ISO/PAS. An ISO/PAS must be reviewed after three years. After six years it must be advanced to an International Standard or withdrawn.

An example for an ISO/PAS is the Java platform of Sun Microsystems.

ISO/TS – ISO Technical Specification

An ISO Technical Specification (ISO/TS) is normative as well. After completion of stage three of the standard development, the committee stage, the document is handed to the TC or the SC. This document is called a committee draft (CD). The TC or SC sends out the CD for vote within a three month period. If two thirds of the P-members vote in favour, the CD is approved as an ISO/TS. An ISO/TS must be reviewed after three years. After six years it must be advanced to an International Standard or withdrawn.

An ISO/TS is a means to make a document official if there is insufficient support for an International Standard or if other existing specifications will get the ISO brand.

An example for an ISO technical specification is the ISO/TS 19103 (Conceptual schema language).

ISO/TR – ISO Technical Report

An ISO Technical Report (ISO/TR) is informative. At any stage after the approval of a New Work Item Proposal (NWIP) the ISO Central-Secretariat may decide to publish a document as an ISO/TR. The ISO/TR requires a simple majority of the P-members of the TC or SC for approval.

The ISO/TR were essentially of three types:
- Type 1 for documents which had been intended to become standards but for which the required levels of agreement could not be attained,
- Type 2 to describe either the directions of standardisation in particular fields or in some instances to make an experimental standard available for trial use, and
- Type 3 is for information only.

In the future, the type 1 and type 2 ISO/TR will be published as an ISO Technical Specification (ISO/TS). The ISO/TR will be retained purely as informative documents, which were the ISO type 3 Technical Reports.

An example is the ISO/TR 19121 (Imagery and gridded data). The ISO/TR 19121 is a type 3 report that basically reviews the de-facto standards in the field of imagery and gridded data.

IWA – International Workshop Agreement

An International Workshop Agreement (IWA) is normative. It is an ISO document that is prepared outside the formal structures of ISO such as the Technical Committees (TCs). The market players, mostly industries, have instead agreed upon the IWA in an open workshop environment in fields, where ISO has no experts or structures available. The IWA is one of ISO's strategies to accelerate its response to market requirements. The option exists to develop the IWA into a full International Standard at a later stage.

1.3.1.5 Template of an ISO standard

All ISO-standards have the same appearance. They are based on the same document lay-out to simplify their creation and usage. ISO provides a template for a standard document. This template is explained in brief in this section.

The ISO 19105 (Conformance and testing) was the first International Standard published of the ISO 19100 family.

An ISO standard document consists of normative and informative pages. The cover sheet, the table of contents, the foreword, and the introduction are called the preliminary elements. The normative pages consist of the scope, the clauses and subclauses. The annexes may be defined as normative or informative.

Cover sheet

The cover sheet contains the title of the document. All ISO 19100 standards have the introductory element "Geographic information" and a main element like "Conformance and Testing".

As French is the second language of ISO, all titles are also given in French, such as "Information géographique – Conformité et essays"

Table of contents

Though the table of contents is an optional preliminary element, it is necessary if it makes the document easier to consult.

Foreword

The foreword makes reference to the ISO directives relevant for this standard and to the Technical Committee responsible for the maintenance.

ISO holds the copyright of all ISO-standards. Reproduction may be subject to royalty payments or a licensing agreement. The copyright notice displays the details.

Introduction

The optional Introduction is an informal description of the content of the standard. It should enable a new reader to evaluate whether the standard is relevant for a given purpose or not.

Scope

The scope defines without ambiguity, the subject of the document and the aspects covered, thereby indicating the limits of applicability.

The scope shall be succinct so that it can be used as a summary for bibliographic purposes.

The scope has to be approved by the plenary of the Technical Committee or Subcommittee in order to align the fields of work of all standards of the family.

Conformance

The "Conformance" clause is only required in some standards in the Information Technology field. This clause shall enable a user to test a product on compliance with the ISO standard.

Normative references, Terms and definitions, Symbols and abbreviated terms

The "Normative references" clause provides a list of the referenced documents cited in the document in such a way as to make them indispensable for the application of the document.

The "Terms and definitions" clause refers to the terminology of the standard. In families of standards like the ISO 19100 it is recommended that one keep track of the terminology of all member standards of the family as a whole.

The "Symbols and abbreviated terms" clause is filled as far as necessary for understanding the standard.

Basic elements

The basic elements contain the provisions of the standard. They are ordered in clauses, subclauses, and paragraphs.

Annexes

Annexes are either normative or informative.

Normative annexes have the same relevance as a chapter in the main body of the standard. While drafting a standard document it is often a matter of style to distribute the basic clauses or the annexes. Often it supports the readability to keep the basic clauses short and to keep major portions of the content in the annexes.

Informative annexes give additional information intended to assist in the understanding or the use of the document.

A complete description of types of clauses possibly being used in an ISO standard can be found at the reference (ISO02 2003).

1.3.2 International Electrotechnical Commission (IEC)

The IEC is the global organization that prepares and publishes International Standards for all electrical, electronic and related technologies. The IEC was founded in 1906. It is a non-governmental organization with its Central Office in Geneva, Switzerland. At present (2003), the IEC has 55 member countries and 10 associated member countries.

IEC and ISO follow almost identical standardisation procedures with standards built on consensus among the interested parties, worldwide solutions, and voluntary initiatives. In general, the IEC uses the same terms for the six stages leading to an International Standard: proposal stage, preparatory stage, committee stage, enquiry stage, approval stage, and publication stage.

1.3.3 ISO/IEC Joint Technical Committee 1 (JTC1)

The ISO and the IEC were founded by the mechanical engineering and electrical engineering communities respectively. Although the separation of the two organizations might have been logical at one time, the era of computers required joint efforts involving both organizations. In 1987, the Joint Technical Committee 1 (JTC1) was created to provide a single, comprehensive standardisation committee that addressed international Information Technology standardisation. The name is confusing in that it implies the existence of other JTC's like JTC2 and JTC3. However, the only Joint Technical Committee in existence is the JTC1.

1.3.4 International Telecommunication Union (ITU)

Founded in 1866, the ITU administers the worldwide use of radio frequencies and satellite orbits. It is the oldest of the three large Standards Developing Organizations. As the ITU is an international organization within the United Nations System (UN) it represents all countries in the world that are presently 189. The Secretary General is located in Geneva, Switzerland.

The ITU is subdivided into three sectors: the Radiotelecommunication Sector (ITU-R), the Telecommunication Standardisation Sector (ITU-T), and the Telecommunication Development Bureau (ITU-D).

The activities of the Telecommunication Standardisation Sector mainly lead to a standardisation for network hardware and network protocols. The work is distributed to a number of study groups. The ITU has recently introduced an "Alternate Approval Procedure" that enables a standardisation procedure as fast as nine months between the start with a stable technical document and the final ballot. The deliverables are recommended standards for international use. As the UN is a treaty organization the recommended standards become immediately national laws in many countries.

1.3.5 Access to standard-documents

All deliverables and additional publications can be bought from the ISO, the IEC, and the ITU. The deliverables of the ISO/IEC JTC1 are available through the ISO- and the IEC-channels.

All standards and other deliverables are available in paper. A growing number of documents can also be ordered as CD-ROM or downloaded as a PDF-file. The online accessed files are watermarked. The shipping of paper documents and CD-ROMs takes a few days only.

The details of the ordering procedure are published on the homepages of the three standards developing organizations. The sections are called ISO Store, IEC Web Store, and ITU publications respectively.

1.4 ISO 9000 family of standards

In non-technical areas ISO is often considered as a synonym for ISO 9000 (Quality Management Systems). The ISO 9000 family of standards sets rules for the efficient structure of the internal workflow of an enterprise. These standards can apply to any type and size of company or institution. Being certified according to ISO 9000 has become an important factor in competition for markets.

Being ISO 9000 certified does not necessarily reflect on the quality of a company's products. It only proves that the internal work flows and their links to customers and suppliers were near optimal when the audit took place. The major challenge to a company remains its success in the market, and is dependent on many factors including expertise of the personnel, sufficient capital, or good service.

Numbering of the ISO 9000 family of standards

There are several individual standards in the ISO 9000 family and, adding to the confusion, the numbering system was changed in the year 2000. Therefore, the formalities are explained in the beginning followed by the concepts of the standard.

The version of ISO 9000 that is presently valid was published in the year 2000. The previous version was published in 1994. The version of an ISO-standard is shown by the year set behind the standard number. For example, ISO 9000:2000.

The ISO 9000:2000 (Quality Management Systems – Fundamentals and Vocabulary) is a basic document for the whole ISO 9000 family of standards.

The ISO 9001:2000 (Quality Management Systems – Requirements) provides a comprehensive list of regulatory requirements needed to assess the ability of a company to meet the customer's satisfaction. This standard is the only one in the ISO 9000 family against which a third-party certification can be carried.

The ISO 9004:2000 (Quality Management Systems – Guidelines for Performance Improvements) provides advice and hints for the continuous improvement of the quality management within a company or institution.

The ISO 19011 (Guidelines on Quality and/or Environmental Management Systems Auditing) is currently (2003) under development. It will provide a guideline to the verification of a system's ability to achieve defined quality objectives. It might be used internally or for auditing suppliers.

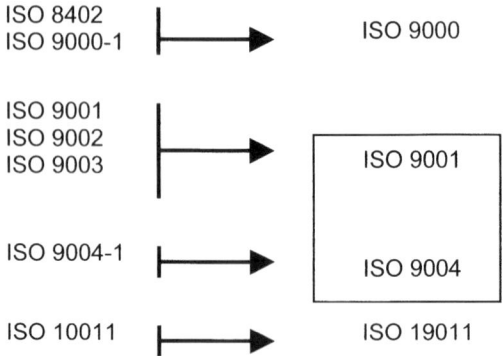

Fig. 1.2. Relation of ISO 9000 standards 1994 edition to ISO 9000 standards 2000 edition
The left side lists the standards of the 1994 edition and the right side lists the standards of the 2000 edition. The arrows indicate, which standards were combined to reduce the overall number of related standards.

Many other ISO standards deal with aspects of quality management. Examples include project management, measuring equipment, manuals, economics, training, and automotive industry. Further details can be found in the bibliography and on the ISO Internet homepage.

A major change from the 1994-edition of the ISO 9000 standards to the 2000-edition is the reduction of the number of individual standards. The new ISO 9001:2000 includes the old ISO 9001:1994, the old ISO 9002:1994, and the old ISO 9003:1994. The diagram 12 explains the relation of the standard numbers of both editions.

ISO Technical Committee 176

The standardisation body responsible for the development of the ISO 9000 family of standards is the ISO/Technical Committee 176 (Quality management and quality assurance). At present, the secretariat is in Canada. In addition to its responsibility for the ISO 9000 family the ISO/TC176 has been entrusted by the ISO/Technical Management Board with the function of consultant to all ISO Technical Committees in the application of quality management and quality assurance. This is to ensure the integrity of the generic quality system standards and to prevent proliferation of sector-specific ISO quality systems standards that may lead to fragmentation of the quality systems of companies in multiple assessments and thus to increased costs. The Technical Management Board is the central technical planning and steering committee of ISO.

Concepts of the quality management system

The implementation of a quality management system at the company level or institution level is described in detail in the ISO 9000 family of standards.

Stage 1: Identify what others expect from the company or institution:
        Customers and end users
        Employees
        Suppliers
        Shareholders
        Society
Stage 2: Determine the processes that are needed to supply products to the customers:
        Customer related processes
        Design and/or development
        Purchasing
        Product and service operations
        Control of measuring and monitoring devices
Stage 3: Plan to close the gaps:
        Identify actions
        Allocate resources
        Assign responsibilities
        Establish a schedule
Stage 4: Carry out the plan:
        Implement the identified actions
        Track the progress to the schedule
Stage 5: Conformance = certification and/or registration

At this stage a company or institution may engage an accredited registration/certification body to perform an audit and certify that the quality management system in place complies with the requirements of ISO 9001:2000. The reasons for a paid independent audit are generally:

- Requirement of a contract with a customer
- Market position
- Regulatory requirements

22   1 Basics of standards

- Risk management
- Internal goals

Unless an audit is performed the results may only be used internally.

The concept of the ISO 9000 family of standards provides a long-term perspective on improvements to the effectiveness and suitability of quality management systems. The ISO 9004:2000 provides the methodology for this continuous effort.

1.5 Cultural and linguistic adaptability

> The language is an essential user requirement:
>
> "Web users stay twice as long and are three times as likely to buy from sites presented in their native language" (Gartner Group).

Technically speaking, interoperability means the ability of heterogeneous datasets to function jointly and to give access to their resources in a reciprocal way. The semantic interoperability includes multi-lingual interoperability. Semantic interoperability is defined as the ability of a user to fully understand the data received in a data exchange in order to be able to make full use of those data if needed.

The ISO/TC211 has recognized this important user requirement by adopting the principles of Cultural and Linguistic Adaptability (CLA) that has been defined as follows by the ISO/IEC JTC 1 (ISOIEC05 1998):

"Cultural and Linguistic Adaptability is the ability of a product, while keeping its portability and interoperability properties,
- to be internationalized, that is, be adapted to the special characteristics of natural languages and the commonly accepted rules for their use, or of cultures in a given geographic region;
- to fully take into account the needs of any category of users."

The term "internationalization" is defined as the process of producing an application platform or application that is easily capable of being localized for (almost) any cultural environment. Examples of characteristics of natural languages are: national characters and associated elements (such as hyphens, dashes, and punctuation marks), writing systems, correct transformation of characters, dates and measures, sorting and searching rules, coding of national entities, (such as country and currency codes), presentation of telephone numbers and keyboard layouts. Related terms are localization, jurisdiction and multilinguism (ISOIEC31 1998).

Cultural and linguistic adaptability should therefore be viewed in a wider sense than only translating an ISO standard to a specific human language.

1.5 Cultural and linguistic adaptability

In the field of metadata, the following strategies are suggested to address cultural and linguistic adaptability requirements. The ISO/TC211 has primarily implemented the methods in ISO 19115 (Metadata) and ISO 19135 (Procedures for registration of geographical information items).

1. The use of an identifier as pivot between linguistic equivalencies

An identifier is defined as "a sequence of characters, capable of uniquely identifying that with which it is associated, within a specified context. A name should not be used as an identifier because it is not linguistically neutral." (ISOIEC33 2003).

The standard ISO 19135 (Procedures for registration of geographic information) will ease the use of identifiers by establishing registers.

The following example shows the identifier "HS:0701" that is used as a pivot between linguistic equivalencies in the Spanish, English, and French language. The Spanish language includes cultural equivalencies between Spain and Mexico:

Table 1.2. Example for cultural and linguistic equivalencies of the word "potato" (ISO06 1998)

Identifier (World Customs organization)	Country code (ISO 3166-1)	Cultural and linguistic equivalencies (3-letter code from ISO 639-2/T)
HS : 0701	484 (Mexico)	(esp) : papa
HS : 0701	724 (Spain)	(esp) : patata
HS : 0701	124 (Canada)	(eng) : potato
		(fra) : pomme de terre

2. The use of code lists instead of free text lists of permissible values

Code lists shall be used instead of free text lists wherever possible.

The example from ISO 19115 illustrates the use of code lists. The metadata element "Maintenance Frequency Code" serves as the pivot between human-language equivalencies.

24   1 Basics of standards

Table 1.3. Code list for "Maintenance Frequency Code"

Name	Domain code	Definition
MD_MaintenanceFrequencyCode	MaintFreqCd	frequency with which modifications and deletions are made to the data after it is first produced
continual	001	data is repeatedly and frequently updated
daily	002	data is updated each day
weekly	003	data is updated on a weekly basis
fortnightly	004	data is updated every two weeks
monthly	005	data is updated each month
quarterly	006	data is updated every three months
biannually	007	data is updated twice each year
annually	008	data is updated every year
asNeeded	009	data is updated as deemed necessary
irregular	008	data is updated in intervals that are uneven in duration
notPlanned	009	there are no plans to update the data
unknown	998	frequency of maintenance for the data is not known

3. Handling human-languages adequately in free text fields.

The standard ISO 19115 (Annex B) provides a method to support multi-languages in free text fields.

The chapter 1, section 5 has been provided by Yves Hudon (Hudon 2003).

1.6 Requirements for future developments of standards

Standards exist to make the life easier, safer, less costly. Industry has speeded up its innovation cycle but the development of standards still follows procedures defined 50 or 100 years ago. Today, many standard users are concerned about the present shape of the universe of international standardisation.

A good standard is the sum of a many individual parts, some of which may be difficult to achieve. The most important properties are the following:

- A standard has to be ready when it is needed
- A standard has to be relevant to the market.
- A standard has to add value to a business and help save money.
- An International Standard has to be applicable worldwide.
- A standard has to be compatible with other standards in the subject area.
- The standards relevant to the subject have to be consistent.

1.6 Requirements for future developments of standards

In order to ensure that the international standardisation process fulfils its intended purpose some members of the user's community are demanding fundamental changes of the structure and the procedures of the standards organizations (Ghiladi 2002)

- The three big Standards Developing Organizations ISO, IEC, and ITU should avoid overlapping work and pool resources like administration, publishing, and marketing.

- The development of new standards must be completed within a maximum of three years in order to keep pace with industry's innovation cycle. The technical experts of the industry should have a stronger influence in the Working Groups, replacing the old fashioned rule of allowing only official representatives.

- The standardisation terminology contains an overwhelming number of structural units, document types, document resources, membership statuses, etc. These terms need to be simplified and harmonized in order to enable the users to more easily understand standardisation abbreviation.

The standardisation of geographic information is heavily influenced by the existing structures and procedures of ISO. Any further work on the geographic information standards should take into consideration the ongoing discussions about fundamental changes of ISO, IEC, and ITU. Recently the maximum time allowed for the development of an ISO standard has been reduced from seven to five years.

2 Geomatics standards

Most readers will be familiar with geographic information systems, or GIS for short, and will think in terms of the GIS with which he or she has gained experience. Some might think in terms of digital maps that are available on the Internet, and others in terms of their areas of responsibility: property cadastre, environmental applications, or fleet management systems. Another important application is disaster management where a GIS can help save lives if rescuers receive an almost immediate and detailed picture of the environment of the disaster location.

This book deals with the standardisation of geomatics. A world-wide standardisation of geomatics is not simple because an enormous number of entirely different applications are served by GIS techniques. For instance, applications can range from city maps delivered on a CD, to global climate monitoring. Geomatics applications run on a single workstation or access thousands of computers connected in a network. Overall, a geographic information system consists of a number of rather independent modules serving data capture, data storage, or data exchange, to mention a few.

The reader might think that some kind of a typical GIS should become an International Standard. This would make the work easier because all "standard GIS" would have a similar shape in core components. Most users demand standardised data exchange formats, because problems with data transfer are well known.

The ISO standards are developed with a long-term perspective. Most of them are written on an abstract level in order to guarantee a long-term stability. In contrast, developments in the geomatics domain continue to make fast progress. Some of the ISO standards for geomatics are implementation-oriented even though this is not the first priority of the ISO/TC211 work.

2.1 History of geomatics standards

Standardisation of computer graphics began in the 1970s but standardisation of geographic information did not begin until approximately 1990. Both efforts originated in Europe.

First generation

The first ISO standard in computer graphics was the ISO 7942 (GKS, Graphical Kernel System). The development began with a workshop on Chateau Seillac in

France in 1976 and the standard was published in 1985. GKS is a generic standard for 2-dimensional vector graphics independent of operating systems, programming languages, and output devices. Language bindings for four important programming languages that followed. GKS has been used in a number of systems and can be considered successful. However, even at the time GKS was published, some of its elements still performed too slowly for larger volumes of mapping data.

During the following years, GKS was extended to include the third dimension. Differing philosophies on the development ISO resulted in two incompatible new standards being published. The European solution, ISO 8805 (GKS-3D), was upwardly compatible to ISO 7942 (GKS) while the US-supported solution, ISO 9592 (PHIGS, Programmer's Hierarchical Interactive Graphic System) was an independent development. GKS-3D and PHIGS share a common core of concepts but they differ in the handling of display lists. GKS-3D supports a linear segment storage concept while PHIGS supports a hierarchical store.

In order to draw a complete picture, two other first generation standards must be mentioned. The ISO 8632 (CGM, Computer Graphics Metafile) and the ISO 9636 (CGI, Computer Graphics Interface). Both of these standards share the same concepts such as primitives and attributes that are currently used in many areas. A direct use of CGM was output but with the advancement of technology, Postscript, PCL, and now SVG would be the currently preferred expression language. This would be the same reason that CGI level interface is now internal to boards.

These standards are no longer used for new developments because some legitimate uses that would commonly be thought of as "Computer Graphics", are not addressed. Amongst these are:

- *Image processing applications*. The successful processing of images was not commonplace before the wide availability of raster devices.
- *Window management*. Graphics standards were developed at a time when the full physical resources on each display would be available to a single application, whereas today, window management systems are based on the dynamic partitioning of a set of resources among tasks.
- *High quality typesetting*. The quality of text facilities available on modern systems far exceed those expected to be widely available in the original timeframe of the standard's development. Consequently, the text facilities are neither simple enough for convenient use by a novice, nor sufficiently sophisticated for control of high quality typesetters (Arnold and Duce 1990).

Graphic format standards

The Working Groups on the computer graphics standards are affiliated to the ISO/IEC JTC1, the "Joint Technical Committee 1" of ISO (International Organization for Standardisation) and IEC (International Electrotechnical Organization). The ISO/IEC JTC1 addresses international Information Technology standardisation that is on the borderline between the traditional ISO and IEC fields of work. The ISO/IEC JTC1 is subdivided into a number of Subcommittees (SCs) whose work has resulted in a large number of standards, most of which are well known, but are rarely related to the work of ISO and IEC. The following table lists the SCs that are relevant for computer graphics and geographic Information Technology and some of the standards they have developed.

ISO/IEC JTC1/SC24: Computer graphics and image processing	GKS, GKS-3D, PHIGS, CGM, CGI, PNG (Portable Network Graphics), VRML (Virtual Reality Modelling Language), BIIF (Basic Image Interchange Format)
ISO/IEC JTC1/SC32: Data management and interchange	SQL/MM (Structured QueryLanguage/ MultiMedia)
ISO/IEC JTC1/SC34: Document description and processing languages	SGML (Standard Generalized Markup L.), HTML (Hypertext Markup Language)

For further details about ISO/IEC JTC1 see chapter 5, section 1.3.

Beginning of the standardisation of geographic information / geomatics

The Europeans have long established traditions in the development of standards for cadastre, cartography, and environmental protection. Following in these traditions, the first project to standardise geographic information started in Europe in 1991. Led by AFNOR, the national standards body of France, the European standardisation organization CEN (Comité Européen de Normalisation) created its Technical Committee 287 (Geographic information). The work of CEN/TC287 resulted in eight European prestandards. The successful conceptual developments were integrated into the larger body of ISO a few years later. The work of CEN/TC287 originally contained a full range of nearly 20 standards. Facing the upcoming establishment of ISO/TC211 in 1994, and to avoid duplicate work, the CEN/TC287 ceased its activities.

The following list identifies the eight European prestandards and the two CEN reports for geographic information. Their titles read as an abstract of the standards later developed by the ISO/TC211 (Geographic information / Geomatics). Many CEN documents were used as draft standards of ISO and many of the CEN experts continued their work in the ISO/TC211.

Some of the above-mentioned prestandards had evolved into CEN standards before they were downgraded to CEN prestandards in order to avoid competition with the ISO/TC211 results.

The demise of CEN/TC287 is not typical of CEN Technical Committees. Since the unification of the continent, the importance of European standards in other subject areas has grown and standardisation is now a priority.

Table 2.1. Prestandards and other deliverables of CEN/TC287

ENV 12009	Reference model
ENV 12160	Data description – spatial schema
ENV 12656	Data description – quality
ENV 12657	Data description – metadata
ENV 12658	Data description – transfer
ENV 12661	Referencing systems – geographic identifiers
ENV 12762	Referencing systems – direct position
prENV 13376	Rules for application schema
CR 13425	Overview
CR 13436	Vocabulary

ENV = European prestandard, prENV = draft European prestandard, CR = CEN Report

The European Union is in the process of developing a Spatial Data Infrastructure (SDI) named INSPIRE (Infrastructure for spatial information in Europe) which is a project of the Environment Directorate-General (DG), the Eurostat DG (statistics) and the Joint Research Centre, the DG for research. A Directorate-General is a central administration level that is directly subordinate to the government of the European Union; the European Commission.

The INSPIRE project is leading to a revival of the CEN/TC287 activities with a plan to develop a European profile of the ISO 19100 standards for geomatics. The formal background for this project is the new edition of a cooperative agreement between ISO and CEN known as the 2001 "Vienna agreement". This agreement enables a close formal cooperation between both standardisation organizations in order to avoid duplicate work. The work on an updated scope of CEN/TC287 is ongoing.

2.2 ISO/TC211

2.2.1 History and work of ISO/TC211

The driving forces behind the establishment of ISO/TC211 were the NATO geomatics Working Group DGIWG and the national standards efforts in the U.S. and Canada. The other two organizations that contributed experiences in the standardisation of geographic information were the International Hydrographic Organization (IHO) and the CEN/TC278 (Geographic Data Files, GDF). The CEN/TC287 (Geographic information) had an established program of work that essentially became the plan for the ISO/TC211 base standards. It was the DGIWG that originally proposed the formation of ISO/TC211, and because it was procedurally easier to have a nation make the proposal, Canada made the proposal in 1994.

Both the original CEN work and the DGIWG work were closer to the implementation level than the current ISO/TC211 standards. Over time, the ISO standards have become more abstract base standards that require profiles and implementation specifications in order to be put to use.

With the establishment of ISO/TC211, a joint worldwide effort to standardise geomatics began. In particular, ISO/TC211 integrated the European and the North American experiences with other regions of the world including Asia, Australia, and South Africa, having since joined the committee.

Scope

The objective of the work of ISO/TC211 is to establish a set of standards for "Geographic information / Geomatics". The standards would specify an infrastructure and the required services for the handling of geographic data including management, acquisition, processing, analysis, access, presentation and transfer. Where possible, the standards would link to other appropriate standards for Information Technology and provide a framework for the development of sector-specific applications (ISO04 2003).

Structure

An ISO Technical Committee consists of the chairman, the members, the Working Groups, some advisory groups, and eventually some Subcommittees.

The chairman, elected by the members of the ISO/TC211, is Olaf Østensen from the Norwegian Mapping Authority.

The members are the national members and liaison members. The *national members* are represented by the authorized standardisation organizations of the respected member country, also called the "national body". Normally the national bodies are *participating members* (P-members) with full voting rights while others are *observing members* (O-members) with an observer (non-voting) status only. The liaison organizations are domain-specific, worldwide or regional organizations that can contribute expertise to the standardisation process. Examples of non-ISO organizations (referred to as *external liaison members)* are the Open GIS Consortium and the International Hydrographic Organization. *Internal liaison members* are other ISO or IEC committees, such as the ISO/TC204 (Intelligent transport systems), that also have interests in geographic information.

Representatives from liaison organizations are invited to take part in the discussions and can receive all information on the standardisation project, but are not granted voting status.

The Working Groups (WGs) form part of the ISO/TC211 hierarchical structure and as the name implies, they are the place where the standards are actually drafted. Until 2001, the ISO/TC211 had the WGs 1-5 and after finishing the work of WGs 1-5 in 2001, the WG 4 remained and WGs 6-9 were created. The WGs combine a number of individual standardisation projects related to a common topic such as the WG 4 (Geospatial services).

The Technical Committee creates advisory groups for special purposes such as the Advisory Groups on Strategy and on Outreach. Such groups analyse the current and future requirements for geomatics standardisation and draft proposals to the Technical Committees on how to proceed. The advisory groups do not belong to the Working Groups; they are directly responsible to the Technical Committee and support the chairman. The Technical Committee has also created a Harmonized Model Maintenance Group (HMMG) that is working on the integration of all partly incompatible models of the ISO 19100 standards.

The ISO/TC211 has no Subcommittees (as of September, 2003).

Statistics

The work of ISO/TC211 can be described in the following numbers (June 2003):

ISO-deliverables
   Total of 40 standardisation projects
   7 International Standards (IS) completed
   11 standardisation projects in the stages Draft International Standard (DIS) or
         Final Draft International Standard (FDIS)
   2 Technical Reports (TR)
   1 Review Summary (RS)
   1 standardisation project withdrawn
   Presently 18 projects still active
   1st International Standard: ISO 19105:2000 (Conformance and testing)
   1st Technical Report: ISO/TR 19121:2000 (Imagery and gridded data)

Members
   29 P-members
   27 O-members
   22 External liaison members
   11 Internal liaison members
   2 Relations to CEN Technical Committees (Europe)
   2 Advisory groups and three other groups
       About 600 individuals involved since 1994

Meetings
   16 Plenary meetings until May, 2003

Program of work

In 1994, the ISO/TC211 started with only 20 standardisation projects. These formed a suite of base standards for geographic information and thus made the Technical Committee a coherent and powerful group. Some experts consider these abstract base standards as the original obligation of the ISO/TC211. The base standards included: the reference model, feature definition, spatial and temporal schema, Coordinate Reference System, portrayal, encoding, quality, and metadata, to mention the most important ones. The majority of the base standards have reached a final stage of development.

The ISO/TC211 is presently undergoing significant changes in order to meet the challenges of the future.

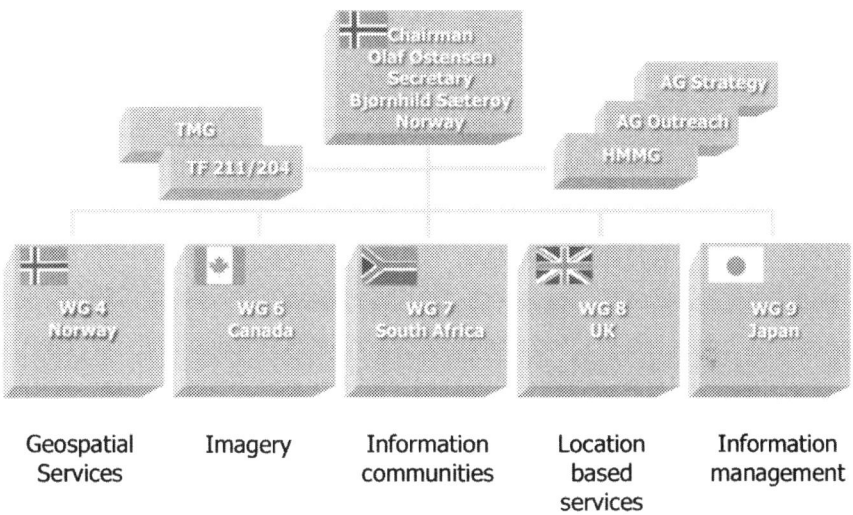

Fig. 2.1. Structure of ISO/TC211 (status 2003) (ISO04 2003)

TMG =   Terminology Maintenance Group
TF 211/204 =   Task Force ISO/TC211 – ISO/TC204
AG Strategy =   Advisory Group on Strategy
AG Outreach =   Advisory Group on Outreach
HMMG =   Harmonized Model Maintenance Group
WG x =   Working Group x

1. Since the completion of the base standards, interest has been focused on their implementation in real world systems. Admittedly, implementation issues are outside the scope of ISO, but they are the only way of proving the viability of the original abstract development. It has been recognised that the field of geographic information is so heterogeneous that the base standards are not always able to fully meet the requirements of the generic and consistent approach that is expected from an ISO suite of standards. Their implementation produces the necessary feedback to optimise the base standards. The cooperation of the Open GIS Consortium is of particular importance for the implementation issues because some of their implementation-type standards, such as the Geography Markup Language (GML), have become drafts for ISO standards.
2. Emerging technologies have started to widen the original scope of ISO/TC211. The industry now expects a tremendous growth in the market for location based

mobile services in the near future and is pushing for standardisation at the boundaries of geographic information, general databases, and mobile positioning systems.
3. The base standards only focused on vector geometries as the field of gridded data – digital images, coverages, and elevation models seemed – too large and unknown to be included in the first attempt at geomatics standards. The projects on imagery standards started in 2001.

The new direction of the ISO/TC211 is formally reflected in the reorganization of the Working Groups (WGs). The only remaining original group is WG 4 (Geospatial services). The titles of the new WGs are WG 6 (Imagery), WG 7 (Information communities), WG 8 (Location based services), and WG 9 (Information management).

As yet, only a few of the standards of the ISO 19100 family have been published as International Standards. However, most of the standardisation projects of ISO/TC211 will be completed in near future. Therefore, all standardisation projects are addressed as "ISO standards" in this book.

Olaf Østensen is Norwegian born in Oslo. He is the chairman of the ISO/TC211 since its beginning and has led the committee successfully in view of the approximately 15 International Standards already completed. In the year 2000 he also became the director of the National Geographic Information Centre, a new unit of the Norwegian Mapping Authority.

During his career Mr. Østensen worked in the Defence Research Establishment and at the Ministry of Environment. He holds a M.Sc. in Mathematics (algebraic geometry) of the University of Oslo. In his research-projects he became familiar with the important programming languages, operating systems, and Internet protocols. He developed and maintained the Norwegian geographic information description and exchange standard (SOSI) for more than a decade.

Since that time Mr. Østensen has always been involved in the work of international standardisation. He was convenor of Working Group 1 (Framework of standardisation) of CEN/TC287 and Norwegian Head of Delegation to the CEN-meetings. He also participated in the standardisation work of ICA (cartography) and DGIWG (military). Recently he took part in the architecture & standards Working Group of INSPIRE, an initiative for a spatial data infrastructure of Europe.

2.2.2 Example of an ISO-compliant GIS

This section provides a basic understanding of the interrelation between the standards of the ISO 19100 family. This understanding is communicated via a small tourism information GIS. This GIS is decomposed from the programmer's perspective in order to relate every component to an individual ISO 19100 standard and to explain which details are standardised and which details are outside the scope of ISO 19100.

Characterisation of the example GIS

The example is a tourism-GIS as shown on the map in figure 2.2. It contains topographic data such as the lake, the woods, and the roads. It also contains the thematic data such as the delineated trails as hiking trails and the opening hours of the theatre. The map is a partial representation of the data and their structure. The map shows only the graphics. The other data might be shown by another type of map or by an interactive program on the computer screen.

The following paragraphs summarise other information that belongs to the GIS "alphanumerically". These data are typical for this kind of a GIS.

A company called "Tourism-GIS-Association" may be responsible for any aspect related to the example GIS.

The GIS data, such as the topographic base map and the info centre of the local community, have different origins. For instance, the geometry, including content information, may have been copied from a topographical base map. The state mapping authority provided additional information about the last revision date and the geometric accuracy of the data. The local community may have supplied the delineation of some of the trails as official hiking trails and the opening times of the campground and the theatre.

From the programmer's point of view, the GIS may be decomposed in the components of data capture, data storage, and data display. In the tourism GIS, the data capture has been completed by introducing the various components from other resources as mentioned above. We can assume that no extra data were surveyed in the field for use in the GIS.

A database will be used for the data storage. The data are not simply lines as shown on the map, but rather grouped into so-called objects like the theatre. Additional data, such as the opening hours, can be linked to this object. Another object is the lake including the shoreline. The objects themselves consist of points, curves, and surfaces and are defined by their coordinates. The mosaic of all objects with the graphic type surface is equal to the complete area covered by the GIS.

A data editing component of the tourism GIS could include a user-accessible function to compute, for example, the approximate lengths of the hiking trails. The user may have to click on the graphics of the trail to cause the system to display the trail's length.

The functionality of the data display component is the shown map. In a general sense, the display is controlled by a reference table that relates every type of object to a graphic representation, such as two parallel lines for the roads.

36   2 Geomatics standards

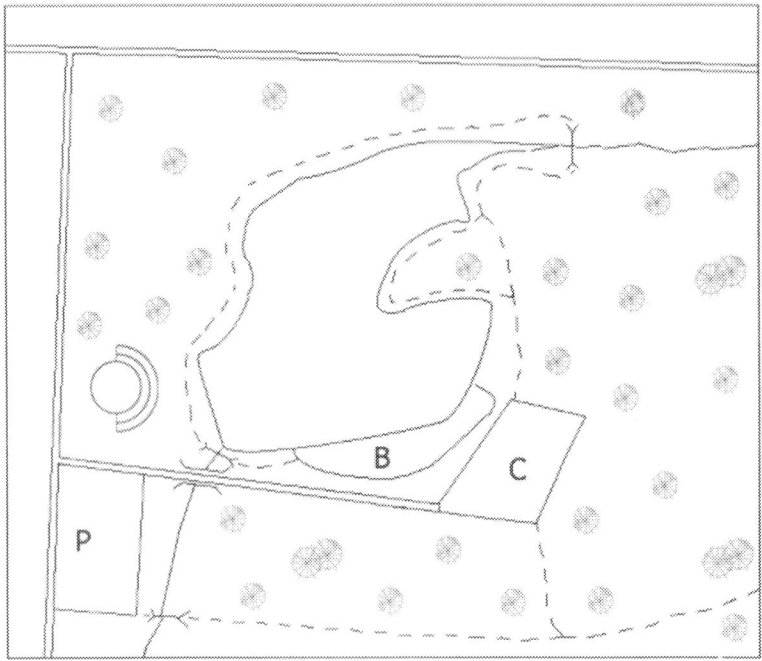

Fig. 2.2. Example of a tourism GIS (B = Beach, C = Campground, P = Parking)

Relation of the ISO 19100 standards

The ISO 19100 family of standards does not form a complete and hierarchical model for the whole universe of geographic information. The standards are more a collection of independent abstract standards for creating and managing Geographic Information Systems. Most standards are only loosely linked.

In the terminology of the ISO 19100 standards, the tourism GIS is an "application schema". The ISO 19101 (Reference model) defines the components of an application schema as shown in figure 2.5.

Data capture

ISO does not standardise the data capture procedures for a GIS. The ISO 19100 standards only provide guidelines and metadata elements to describe the origin and the quality of the data. The relevant standards are ISO 19113 (Quality principles), ISO 19114 (Quality evaluation procedures) and ISO 19115 (Metadata).

The following paragraph provides a detailed relation of the tourism GIS to the quality concept of the ISO 19100 standards. Let us assume that we have to write a "quality evaluation report" about our GIS. This report is a standardised procedure of

ISO 19114. It might report values of the "data quality elements" in the following way:
- Completeness: 100% coverage
- Logical consistency: no errors found
- Positional accuracy: ± 5m
- Temporal accuracy: last map revision 3 years ago, community information supplied last spring
- Thematic accuracy: topographic base map does not supply shelters and other similar elements for tourist purposes

It might also report values of the "data quality overview elements", including:
- Purpose: tourism-GIS will enable an online information source of tourist details and it will also promote tourism in the area
- Usage: dataset was assembled for the tourism-GIS
- Lineage: dataset was created under the supervision of the "Tourism-GIS-Association"

The ISO 19114 (Quality evaluation procedures) provides advice on different levels of detail for the quality check. For example, a "direct evaluation" includes a report on all "data quality elements". An "indirect evaluation" includes only the "data quality overview elements".

Let us assume we have only checked the purpose, usage, and lineage, and that we will write the results in a report. The ISO 19115 (Metadata) provides the formal names of all the necessary data elements. Accordingly, the "evaluationMethodType" of the report will be "indirect" in this case.

Conceptually, the ISO 19115 covers the complete list of the metadata elements of all ISO 19100 standards. An amendment to ISO 19115, ISO 19115:2 (Metadata for imagery), will integrate further elements related to imagery. The detailed explanation of every element within its thematic context is placed in each individual standard of the 19100 family.

Data storage

In the terminology of the ISO 19100 standards, the tourism-GIS is referred to as a *dataset*. The elements of the dataset were called "objects" in the beginning of this section. However, although the term "object" is familiar in the context of object-oriented languages, in the domain of geographic information - the term "feature" is more widely used to denote an element of the dataset. The term "feature" has also been adopted by the ISO 19100 standards.

The ISO 19109 (Rules for application schema) contains all the definitions related to features. The core of ISO 19109 is the "General feature model" (GFM). It states that a feature may have attributes and operations. Attributes identify whether a feature is a point, a curve, or a surface and operations identify whether the feature changes according to external influences, like the theatre being closed in winter. Features may be logically grouped into larger units. For example, woods and fields both belong to land use. A single element in the dataset, such as one of the hiking trails, is called an instance of the feature "hiking trails". Some features have special relationships that become important once a map is drawn, or networking algorithms are applied to the dataset. An example for this case is the bridge crossing the streams. The bridge will always go over the stream at a different elevation than the water surface. This relationship is an association between features.

It is obvious that the possible features of a GIS have to be agreed upon before the population of the database starts. The result of this design process is basically a listing of all allowed features together with their attributes, operations, and associations. The listing is called a feature catalogue. The complete result is called a *data model*.

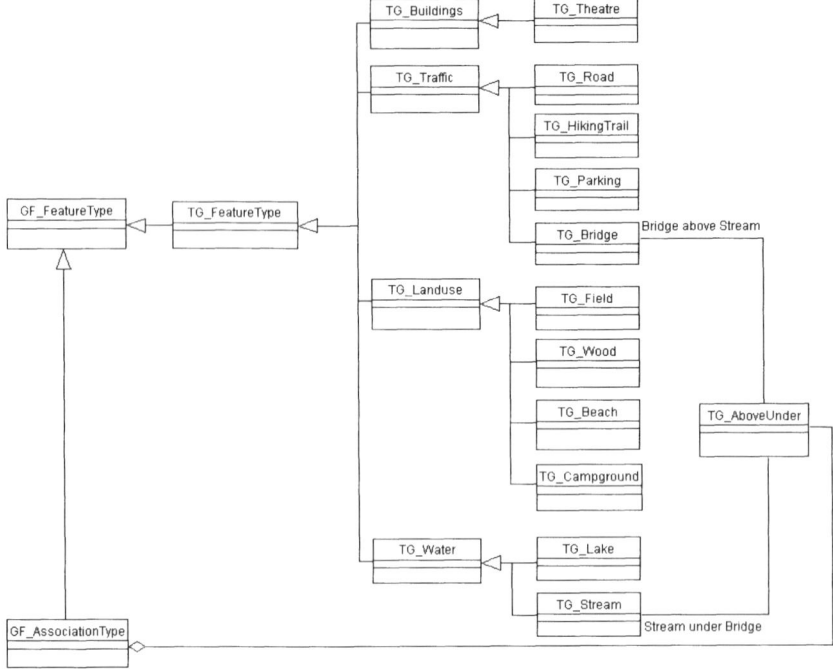

Fig. 2.3. Data model of the tourism-GIS shown as a UML class-diagram
All classes of the tourism-GIS start with the two characters "TG" to indicate their relation to the tourism-GIS.
(Hint: UML, the Unified Modelling Language, is addressed in section 3.1.1)

The ISO does not standardise feature catalogues and data models. The only guidance provided is the ISO 19110 (Methodology for feature cataloguing). This standard provides the general rules and a catalogue-template to support a complete and consistent listing.

The following list is a simplified feature catalogue for the tourism-GIS.

Table 2.2. Feature catalogue of a tourism-GIS

GF_FeatureType high level	GF_FeatureType low level	GF_Operation	GF_AttributeType	GF_AssociationType
Buildings	Theatre	closed in winter	GM_Point	
Traffic	Road		GM_LineString	
Traffic	Hiking trail		GM_LineString	
Traffic	Parking		GM_Polygon	
Traffic	Bridge		GM_Point	bridge over stream
Landuse	Field		GM_Polygon	
Landuse	Wood		GM_Polygon	
Landuse	Beach		GM_Polygon	
Landuse	Campground	closed in winter	GM_Polygon	
Water	Lake		GM_PolynomialSpline	
Water	Stream		GM_PolynomialSpline	stream under bridge

It was stated earlier that the attributes of a feature could be of the type point, line, or area. In fact, this was an over simplification. The ISO 19100 standards provide a great variety of geometry classes and rules as to how they are related to each other. The most comprehensive collection of geometry classes can be found in the ISO 19107 (Spatial schema).

The tourism-GIS uses four geometry classes of ISO 19107: GM_Point, GM_PolynomialSpline, GM_LineString, and GM_Polygon. The class GM_Polygon denotes surface-type features.

The ISO 19107 distinguishes between primitives and complexes. Primitives are geometries that *do not* include their end-points. Complexes are geometries that *do* include their end-points. If a topological network underlies our tourism-GIS, the geometries must be connected at their end-points. Thus, the geometry consists of complexes.

The ISO 19100 family does not standardise the way the geometry and the topology are handled by the database. It does not standardise data types nor details of the topology. Advanced systems only store one point at one position allowing a fast search of the neighbouring features. Simpler systems only store individual geometries and allow multiple points at a position.

All geographic positions of the data are defined according to a Coordinate Reference System. The ISO 19111 (Spatial referencing by coordinates) sets the rules for the definition of Coordinate Reference Systems. In the case of our tourism-GIS, a local Coordinate Reference System is a sufficient solution. A local Coordinate Reference System is an *engineering* Coordinate Reference System according to the ISO 19111. To simplify the usage of coordinates it is advisable to avoid negative values. For this purpose the complete definition area of the GIS must lie in the first quadrant, or seen from the other perspective, the origin of the Coordinate Reference System must be beyond the west and south of the definition area. The x- and y-axis might point to east and north respectively. The terms "datum" (= origin) and the "prime meridian" (= axes to north) provide a local reference only. The ISO 19111 does not standardise any parameter describing global or local Coordinate Reference Systems.

Descriptive parameters are kept independent of the ISO 19100 family of standards. In order to guarantee a standardised use of important parameters, ISO 19100 provides one or more registries. The rules for creating these registries are described in the ISO 19135 (Procedures for registration of geographical information items). A specific registry for Coordinate Reference Systems is defined in the ISO 19127 (Geodetic codes and parameters). If parameters are needed, they can be requested from one of the ISO-approved registration authorities.

Data display

According to the ISO 19117 (Portrayal), the graphic representation is handled independently from the feature data. In the case of the tourism-GIS an individual feature portrayal is defined for each of the features such as the roads and hiking trails. All necessary feature portrayals are summarised in a listing called the *feature catalogue* that contains one entry for each feature in the dataset. Consequently, the feature catalogue of the tourism-GIS has 11 entries.

Each feature portrayal points to a rule that will be applied when drawing that feature. The rule itself is outside the scope of the ISO 19100 standards. In order to handle more sophisticated graphics, like the automatic opening of a road intersection, the ISO 19117 allows the usage of external functions. The summary of all portrayal elements is the portrayal specification. During the output of the feature data, this specification is applied to the data stream in order to generate the cartographic representation.

The following table shows the portrayal catalogue of the tourist GIS (class PF_PortrayalCatalogue)

Table 2.3. Portrayal catalogue of a tourism-GIS

PF_FeaturePortrayal	PF_PortrayalRule
Beach	white area, border with black solid line, thickness 0.2 mm, letter B at the representativePoint
Bridge	place bridge symbol
Campground	white area, border with black solid line, thickness 0.2 mm, letter C at the representativePoint
Field	white area, border with black solid line, thickness 0.2 mm
Hiking trail	dashed black line, line 2 mm, space 2 mm, thickness 0.2 mm
Lake	white area, border with solid black line, thickness 0.2 mm
Parking	white area, border with black solid line, thickness 0.2 mm, letter P at the representativePoint
Road	two black solid parallel lines with a thickness of 0.2 mm each, one line 1 mm to the left and the other line 1 mm to the right of the road's axis, interrupt lines to open roads at intersections,
	show roads at highest priority if overlay with other graphics occur
Stream	solid black line, thickness 0.2 mm
Theatre	one circle, and two concentric semi-circles, solid black lines, thickness 0.2mm
Wood	white area, border with black solid line, thickness 0.2 mm, place tree symbols at irregular positions

Data exchange

ISO does not standardise any exchange format but data exchange is addressed in the ISO 19118 (Encoding). It recommends that an exchange format be built upon the currently widely used Extensible Markup Language (XML). Also, ISO does not deal with the technical details of transferring data between different hardware platforms. The ISO 19118 only states that the application schema of the two systems between which the data exchange takes place, must be alike. This simply means that a system to which the tourism-GIS data may be transferred, has to provide the same feature types: lake, wood, and beach, etc., and the same structural information. If the feature types do not match, information is lost during the transfer. If, for example, in the second system only one feature for traffic is available, then roads and hiking trails would become the same feature and could not be distinguished from one another.

2.2.3 Reference model (ISO 19101)

Among geomatics specialists a consensus has emerged that the field of geographic information is a specialisation within the Information Technology field. At the same time, there is a growing recognition among users of Information Technology that indexing by location is a fundamental way of organising and using digital data. To

David McKellar is Canadian. He was born in Glasgow, Scotland, and immigrated to Canada in 1957. He has joined the Department of National Defence (DND) 25 years ago. His current position is with J2 Geomatics, Imagery & Counter Intelligence.

Mr. McKellar is said to be the father of the ISO/TC211 as he conceived and planned the establishment of this committee, thus serving the requirements of the international geomatics community, and the interests of Canada as well as of the geomatics group of NATO (DGIWG).

Mr. McKellar received Bachelors and Masters degrees in the Geomatics field from the Ryerson University in Toronto, Ontario, and the Ohio State University in Columbus, Ohio, respectively. In three of his past positions Mr. McKellar served as the President of the Canadian Institute of Geomatics, as the chair of the Canadian General Standards Board (CGSB) Committee on Geomatics, and as Head of the Canadian Delegation to ISO/TC 211 (Geographic Information/Geomatics).

Mr. McKellar has also been very active in the community for many years and is a Past President of the Rotary Club of West Ottawa.

meet these needs, standardisation of geographic information in the ISO 19100 series is based on the integration of the concepts of geographic information with those of Information Technology. Whenever possible the development of standards for geographic information considers the adoption or adaptation of generic Information Technology. It is only when this cannot be done that geographic information standards have been developed. The basic principles of Information Technology can be found in standards of ISO and IEC.

Viewpoints

The usage of viewpoints is a common method of decomposing large distributed software systems during the design process. This core-model is standardised as ISO/IEC 10746 (Information Technology — Open Distributed Processing — Reference Model). It is also the foundation of the reference model for the ISO 19100 family of standards.

The ISO/IEC 10746 describes five viewpoints:

1. The **enterprise viewpoint** is concerned with the purpose, scope and policies of an organization in relation to geographic information systems. This viewpoint is used to generate requirements and varies among different organizations and, therefore, it is not within the purview of the ISO 19100 family of standards.

2. The **information viewpoint** is concerned with the semantics of information and information processing. A specification developed from this viewpoint provides a model of the information in a GIS and defines the processing that is performed by such a system. The information provides a consistent common view on information that can be referenced in a GIS. The information viewpoint is the most important viewpoint for the ISO 19100 family of standards.

3. The **computational viewpoint** is concerned with the patterns of interaction among services that are part of a larger system. A service may be the interaction between the system and the client, like the window-interface on the screen, or between a set of other services such as the data retrieval from a database in the background. The computational viewpoint is the second most important viewpoint for the ISO 19100 family of standards.

4. The **engineering viewpoint** is concerned with the design of implementations within distributed, networked, computing systems that support system distribution. As the implementation is not the focus of the ISO 19100 standards, there is little emphasis placed on this viewpoint.

5. The **technology viewpoint** is concerned with the provision of the underlying hardware and software infrastructure within which services operate. Again, as this is out of scope of the ISO 19100 family of standards, there is little emphasis placed on this viewpoint.

Figure 2.4 displays the viewpoints graphically.

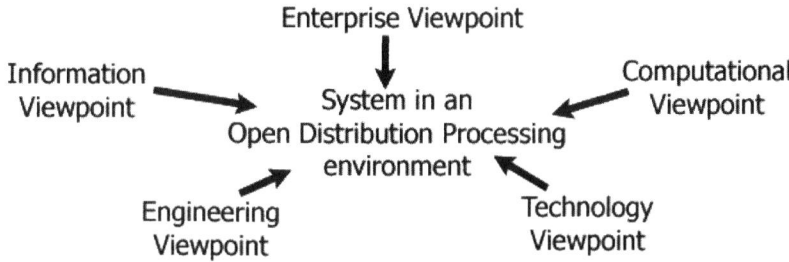

Fig. 2.4. Viewpoints of the Open Distributed Processing — Reference Model

The ISO 19101 (Reference model) defines a hierarchically structured reference model for the ISO 19100 family of standards. Primarily, this reference model is a special application of the "information viewpoint" which is identified as the most important one for the standardisation of geomatics. One component of the reference model is based on the "computational viewpoint" that addresses the services.

The majority of individual standards of the 19100 family are considered as services that operate on datasets. The services are discussed in detail in section 3.2.1 "Services".

The ISO 19101 is primarily for standard developers. However, understanding this standard is also helpful for users of the other geomatics standards because it leads to an understanding of the rationale of the ISO 19100 series.

Conceptual Schema Modelling Facilities

The background for structuring the reference model of geographic information is a number of principles on which Information Technology has agreed upon. The ISO/IEC 14481 (Conceptual Schema Modelling Facilities (CSMF)) lays out a schema for the design of computer software that has four levels of abstraction.

The *meta-meta model level* is the highest level. Though the name is somewhat obscure, the meaning of the meta-meta model is simple. It is the description of a software system with natural language. The meta-meta model level is not a matter of standardisation.

The *meta model level* is below the meta-meta model level in the hierarchy. It contains the description of the concepts of a software system with a formalised language that, in the case of the ISO 19100 standards, is the "Unified Modelling Language" (UML). The concepts include terminology, operations, and assumptions needed to construct applications. Typical examples of meta models are the UML diagrams found in this book and throughout the ISO 19100 standards.

The *application model level* exists below the meta model level. An application model contains all detailed definitions needed to tailor a software system to a specific application. In the case of geographic information, this might include the definition of feature types like roads and buildings and their possible attributes and operations. The documentation of an application model is called an application schema.

The *data model level* contains the datasets and is the lowest level. In the case of geographic information, the datasets are composed of one or many features including their attributes and positions. The datasets are the actual data whereas the application schema only sets the frame for the possible features.

Meta-meta model and meta model

Any consideration on modelling starts with the real world. The chosen piece of the real world that one wishes to describe in a model is known as a universe of discourse. A first result of the modelling process is a conceptual model that describes and limits the universe of discourse. A model is an abstract and represents only a part of the real world. An example of the real world could be the landscape depicted in the map in figure 2.2. A conceptual model is the list of features relevant for the tourism-GIS in table 2.2.

A conceptual schema language provides the semantic and syntactic elements used to rigorously describe the conceptual model and to convey consistent meaning. A conceptual model described using a conceptual schema language is called a conceptual schema. A conceptual schema of the tourism-GIS is shown in figure 2.3. The conceptual schema language prescribed in the ISO 19100 series is UML. The feature types that are shown represent the semantics. The lines specifying the exact type of associations are the syntactic elements. The conceptual schema language is addressed in more detail in section 3.1.1.

Application model

The application model of a GIS completely defines its data elements (classes) and their interrelation. The heart of the application model is a detailed description of the data structures. This description is called an application schema. The tools to describe the data structure are the feature catalogue and the General Feature Model (GFM). The feature catalogue is a formal list of the feature types that are determined as relevant and qualified for the application. The General Feature Model sets the frame for the definition of every feature type. An example of a feature catalogue is provided in section 3.4.2. The General Feature Model is explained in detail in section 3.2.2. Both tools have dedicated ISO 19100 standards.

An application schema also consists of components that are relevant to all feature types. These components are the Coordinate Reference System and data quality describing elements. The Coordinate Reference System is explained in detail in section 3.2.5 and the quality aspects are addressed in section 3.2.9. Both components have their own standards of the ISO 19100 family.

Data model

The data model contains the data being collected for an application and the services that operate on the data. The data model includes all features and their describing data that make up the complete GIS.

The data consist of the dataset and the metadata.

The dataset includes only the features and their positions.

The metadata is the describing data. The ISO 19115 (Metadata) gives a good guideline for the selection of metadata. The ISO 19115 provides a standardised list of metadata elements. The complete list of those elements is given in the annex B of this book. Further describing elements that are not available in the ISO 19115 may be modelled as metadata elements beyond the scope of ISO 19115 or as attributes of the feature classes.

The programs that use the dataset and the metadata to serve a certain purpose for the user are called geographic information services. In the ISO 19100 standards it is stated that services operate on the dataset and reference additional information from the metadata. Most of the ISO 19100 standards define such services.

Conceptual modelling and domain reference model

Based on the Conceptual Schema Modelling Facilities, the ISO 19101 sets forth the process for conceptual modelling, the domain reference model, the architectural reference model, and profiles.

The conceptual modelling covers the meta model. According to ISO/IEC 14481, a number of principles govern the use of conceptual modelling and the development of conceptual schemas. The most important principles are listed below:

1. The **100% principle** states that all (i.e.; 100%) of the relevant structural and behavioural rules about the universe of discourse shall be described in a conceptual schema. Thus, the conceptual schema defines the universe of discourse.
2. The **Conceptualisation principle** states that a conceptual schema should contain only those structural and behavioural aspects that are relevant to the universe of discourse. All aspects of the physical external or internal data representation should be excluded.
3. The **Self-description principle** states that normative constructs defined in the standards and profiles of the ISO 19100 series shall be capable of self-description.

The domain reference model covers the application model and the data model. The details of domain reference model have been explained above.

The figure 2.5 summarizes the relations between the Meta-model, the Application-model and the Data-model.

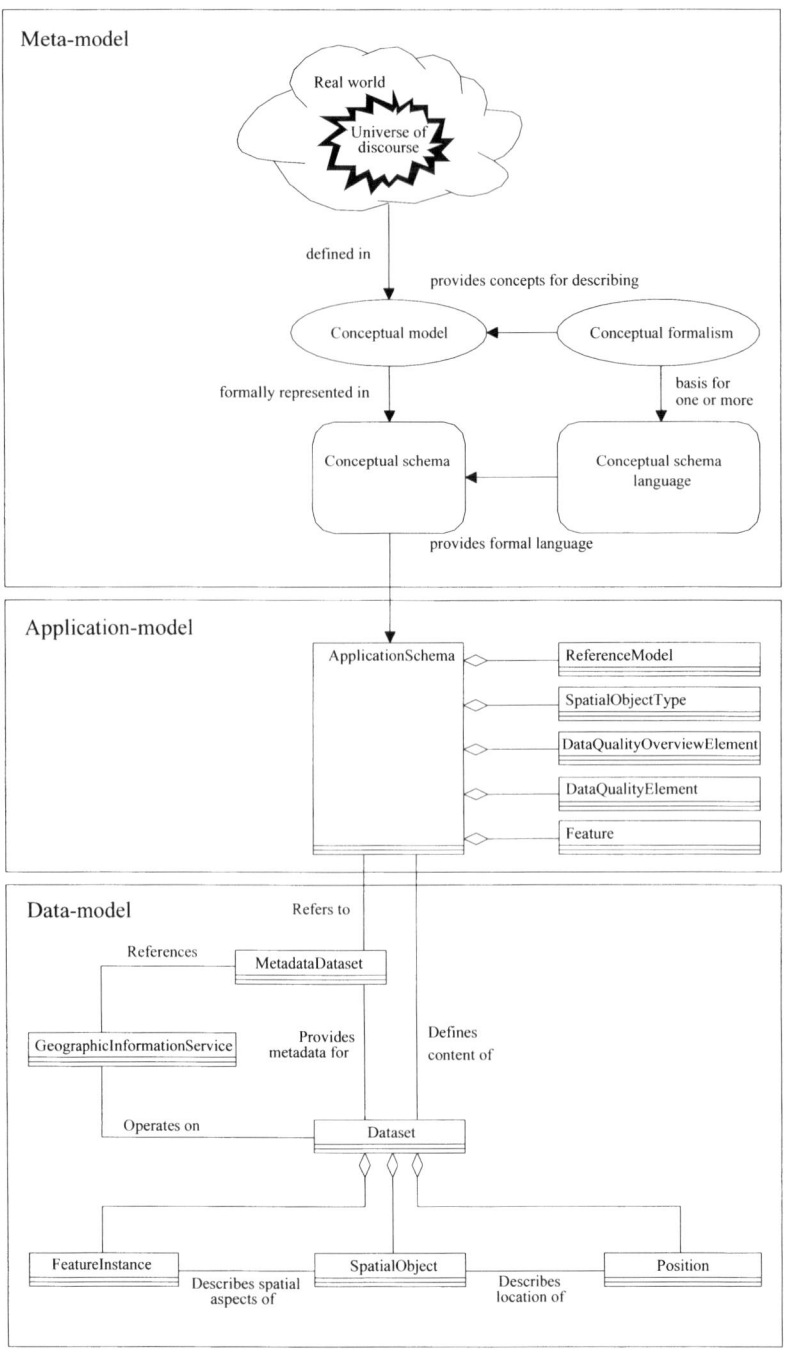

Fig. 2.5. Conceptual modelling and domain reference model (ISO34 2002)

Architectural reference model

The term "architecture" is used here in the context of service architecture. The architectural reference model lays out a concept for structuring the great number of geographic information services involved. The basis for the architectural reference model is the ISO/IEC Open Systems Environment Reference Model (OSE-RM), described in ISO/IEC TR 14252. This standard completely addresses the services for Information Technology. The ISO 19101 standard defines an <u>Extended</u> Open Systems Environment (EOSE) Reference Model that includes the specific services for geomatics. Figure 3.9 provides a complete view of the IT-services and the geographic information services.

Profiles

A profile of a standard defines a subset of the elements of the standard in order to meet specific needs and to eventually avoid a heavy overload of functionality. Profiles of the ISO 19100 standards may combine subsets of different standards - ISO 19100 family standards and others. A special case of future profiles are functional

Hans Knoop was born in Magdeburg in the eastern part of Germany. He is the former Head of the 'Division of Surveying and Cadastral Affairs – Cadastre, GIS, International Cooperation' of the Federal State of Niedersachsen with 3.500 staff. He is also Head of the Department of 'Surveying and Geographic Information' of DIN, the German Institute for Standardisation in Berlin.

Dr. Knoop holds a Diplom-Ingenieur and a Doktor degree, both of the Technical University of Hannover. With his studies in geodesy and his research efforts in electronic tacheometry he contributed one of first developments of computer based surveying techniques. Later Dr. Knoop became a professor for Cadastral Surveying and GIS at the Technical University of Braunschweig, Germany, and an appointed honorary professor of Wuhan Technical University in China.

Dr. Knoop belongs to the pioneers of international standardisation of GIS. From the beginning he was a member of CEN/TC287, the European GIS standardisation, and later member of ISO/TC211. His efforts focused on networking the ISO-standards with the developments of neighbouring subjects such as surveying (FIG), photogrammetry/remote sensing (ISPRS), and education. He is a member of the Advisory Group on Strategy and also co-chairs the Advisory Group on Outreach.

standards that have been developed outside the world of the ISO 19100 standards before the work of ISO/TC211 had started. These functional standards are planned to become profiles of the ISO 19100 family.

Profiles and functional standards are discussed in section 3.1.4.

2.3 Future of geomatics standardisation

2.3.1 Implementation strategies

Since the development of the basic standards of the ISO 19100 family has almost come to completion their implementation has started. The strategies for implementing the mostly abstract ISO 19100 standards is not a standard on its own. However, some typical approaches can be reported.

The ISO/TC211 has about 60 member countries and 20 external liaison organizations. While participating in the work of ISO/TC211 the members have committed themselves to adopt the resulting standards in their home environment. However, the implementations will vary from country to country, from organization to organization, in some cases even from province to province.

Hardly any implementation will require the complete functionality of the ISO 19100 series. Instead, mostly only a subset of the abstract standards will be substantiated. The implementations remain compatible as far as their application schema overlap. The usage of the ISO 19100 guarantees a common underlying abstract model that enable compatibility if the applications schemas are the same.

The Open GIS Consortium (OGC) plays an important role on the implementation level. In theory, the ISO/TC211 develops the abstract standards and the OGC developes the implementations standards. In practice, the borderline is not drawn that clearly.

The ISO/TC211 has created some of the ISO 19100 standards based on proposals from the OGC. Two typical examples are the ISO 19123 (Schema for coverage geometry and functions) and the ISO 19133 (Location based services tracking and navigation). Those standards are implementation standards. On the other hand the OGC has developed a comprehensive Abstract Specification that has been completed in 1999 but hardly touched since.

The partitioning of the work between the ISO/TC211 and the OGC has turned out to be fruitful for the geomatics community. The development of an ISO standard may take up to five years. This is acceptable for long-term abstract standards that enjoy a consensus among the large international community. But that time is far too long for implemented computer software. Programmed tools would just ignore the official standards. The implementation became the domain of the OGC. An example is the Geography Markup Language, version 3.0 (GML 3.0). The GML 3.0 is an almost complete implementation of the ISO 19107 (Spatial schema). It has been questioned whether GML 3.0 is needed to become an ISO standard too. Presently the ISO project 19136 (Geography Markup Language) is still under development.

Many member countries and liaison organizations of ISO/TC211 will adopt the implementation of the ISO 19100 standards from the OGC.

2.3.2 Ongoing work of ISO/TC211

Though the major standards of the ISO 19100 family have been completed the work of the ISO/TC211 is ongoing. The present projects focus on the implementation of the completed work or address new fields such as the Location Based Mobile Services.

The standardisation projects mentioned in this chapter are implementation-oriented and in the early stage of a New Work Item Proposal (Spring 2003).

ISO 19101-2

The ISO 19101-2 (Imagery reference model) extends the reference model of ISO 19101 towards geographic imagery and gridded data applications. Imagery and gridded data are the dominant form of geographic information in terms of volume. However, ISO standardisation of geomatics focused on the world of vector data in the first phase and thus the necessity to rethink the models already created. A similar update of an existing standard takes place with the ISO 19115 (Metadata).

ISO 19115-2

The ISO 19115-2 (Metadata – Part 2: Metadata for imagery and gridded data) is a complementary standard to the existing ISO 19115 (Metadata) and will define metadata elements to support imagery, and gridded data. It will support the collection and processing of natural and synthetic imagery and their derived products. To permit the development of ISO 19115 to proceed, inclusion of metadata definitions for imagery and gridded data was deferred until the framework for these data was more fully specified.

ISO 19137

The ISO 19137 (Generally used profiles of the spatial schema and of similar important other schemas) is in the early stage of a New Work Item Proposal. This future standard shall support the creation of profiles and application schemas in particular of the ISO 19107 (Spatial schema), the ISO 19108 (Temporal schema), the ISO 19111 (Spatial referencing by coordinates), and the ISO 19118 (Encoding). Those profiles and application schemas shall integrate existing standards and de-facto standards such as GeoVRML, ALKIS (Germany), Interlis (Switzerland), and DIGEST (NATO) into the world of ISO 19100 standards.

ISO 19139

The ISO 19139 (Metadata – implementation specifications) aims at defining an XML encoding for the metadata elements defined in the ISO 19115 (Metadata). Because of the abstract nature of the ISO 19115 specification, the actual execution of geographic information metadata could vary based on the interpretation of the metadata producers. The ISO 19139 is meant to enhance interoperability by providing a common specification for describing, validating and exchanging metadata about geographic datasets.

ISO 19140

Within the ISO/TC211 the discussion has started on methods to organize the necessary continuous update of the ISO 19100 standards. A first attempt is the New Work Item Proposal ISO 19140 (Technical Amendment to the 191** Geographic information series of standards for harmonization and enhancements). Presently the organizational shape of the effort is not yet clear. The discussion about the number of involved expert teams, the time frame, and the responsibilities has not come to a conclusion.

2.3.3 Directions and Activities

The work of ISO/TC211 (Geographic information / Geomatics) has completed its first lifecycle. Almost all of the base standards have reached or passed stage five, the enquiry stage. The documents are Draft International Standard (DIS), Final Draft International Standard (FDIS), or International Standard (IS).

Trends

Though the future direction of ISO/TC211 is partly open, the prevailing trends are clear. The base standards will move to practical implementations with responsibilities devolving to industry and national or supranational government institutions. The ISO/TC211 member countries are committed to adopt the ISO 19100 standards. The support facilities for implementers of the ISO/TC211 are limited.

There will be a strong case for using profiles and application schemas of the ISO 19100 family to build well-tailored environments for specific fields of applications. One example of these fields is the property cadastre.

The ISO/TC211 will also be shifting its focus to neighbouring areas of work where the industry has expressed a need for standardisation. The most prominent example of a new area is the location based mobile services.

Strength

The ISO/TC211 has become the worldwide-accepted umbrella group for the standardisation of geomatics. It has succeeded in engaging the national bodies and kindred organizations in the development of the suite of International Standards. The ISO/TC211 has evolved as the competent group that is able to represent geomatics, fairly independently of national and industrial influences and with a sufficiently clear border to other fields of ISO.

Weakness

In geomatics a lot of private standards are already in place with some of them being de-facto standards. The newer ISO standards cannot replace them on a short-term basis and it might therefore, take a long period until the ISO standards are implemented.

Opportunities

The ISO/TC211 has formulated the abstract basis for the whole world of geomatics. Through its neutral position it has become the co-ordinator of many activities in the geomatics business. Productive collaborations among liaison and national members have been or will be initiated. The first co-operative agreement was signed with the Open GIS Consortium (OGC), followed by the International Hydrographic Organization (IHO). Other organizations such as the Digital Geographic Information Working Group (DGIWG) are likely to follow too. The close cooperation with national associations is also anticipated.

The registry (ISO 19135) opens a flexible interface from the ISO 19100 standards to a wide range of worldwide applications. The registry will not only allow keeping track of ISO compliant standards but it will also be a continuously growing source of geomatics implementations beyond the scope of ISO/TC211.

The ISO 19100 standards were chosen as the reference for a number of emerging technologies.

The Global Spatial Data Infrastructure (GSDI) is promoting a globally coordinated approach to geomatics.

The Infrastructure for Spatial Information in Europe Initiative (INSPIRE) sets the framework for a geomatics infrastructure in Europe.

The United Nations (UN) are going to build their implementations in geomatics on ISO 19100 standards.

The ISO/TC211 has established an Advisory Group on Strategy to address future directions and report to the chairman.

Promotional activities

In the past there were concerns about the ISO 19100 standards not being well known and as a consequence not widely used. The ISO/TC211 has therefore started an outreach campaign to create awareness of its standards.

The creation of awareness will support application-oriented activities including implementation of transfer standards by vendors, implementation of data cataloguing standards by data producers, and implementation of metadata standards by vendors and general users.

The development of awareness will include education and training as well as the creation of user communities.

The ISO/TC211 has established the Advisory Group on Outreach to follow up the promotional activities.

Political background

Many countries are beginning to understand the tremendous value of geographic information systems and their data for governing the commonwealth. In the USA, the homeland security takes a strong interest in building a geographic data infrastructure for improving the emergency management including the fight against terrorism and better disaster management. The U.S. also supports the development of a global spatial data infrastructure (GSDI) because GIS is believed to offer an effective weapon against global terrorism.

The development of the European spatial data infrastructure is governed by the long tradition and high quality of the property cadastre and topographic portrayal on this continent. Another strong component is led by the needs for environmental protection. The INSPIRE-initiative is taking care of an integrated approach of all countries of the European Union.

2.4 Roadmap to the ISO 19100 standards

The figure 2.6 shows a roadmap to all ISO 19100 standards and points to the chapters of this book.

The roadmap shows the way of structuring the ISO 19100 standards that is used in this book. The top-categories – infrastructure, basic, imagery, catalogue, and implementation standards – are not official terms. However, they are often used in internal discussions and they are helpful to understand the overall structure of the ISO 19100 family.

The basic standards designate those standards that belonged to the original scope of ISO/TC211 and that do not apply to all other standards like the infrastructure standards. The basic standards must not be confused with the base standards which have the property to enable the derivation of profiles.

The georeference category is unofficial too.

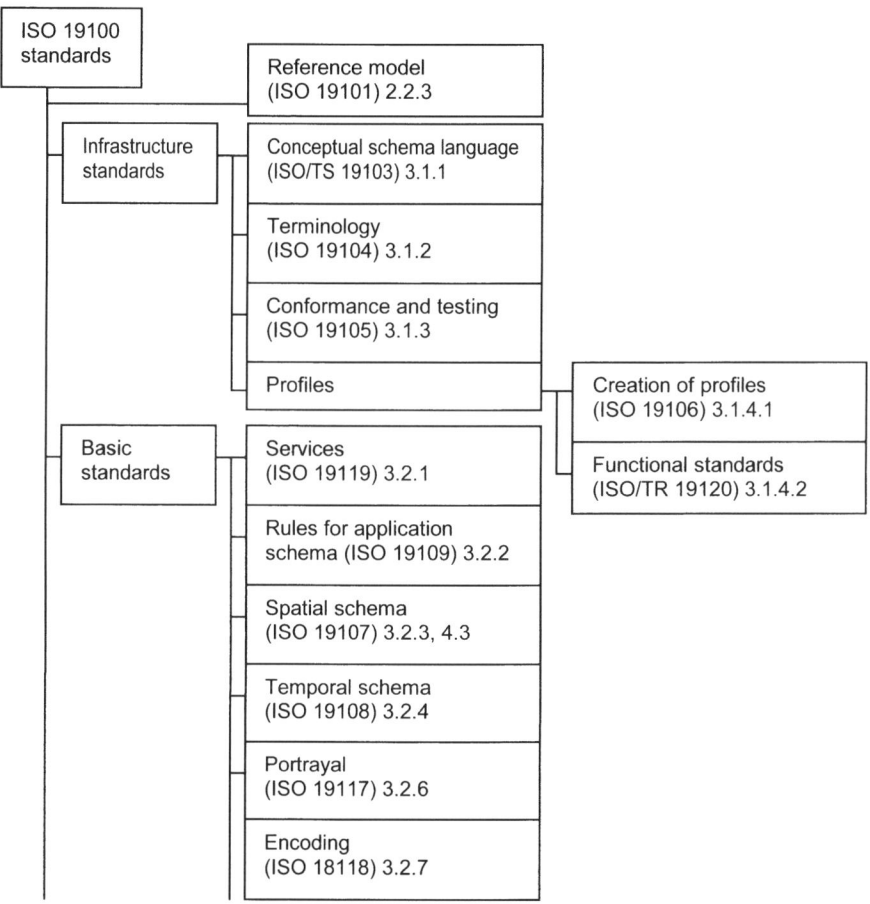

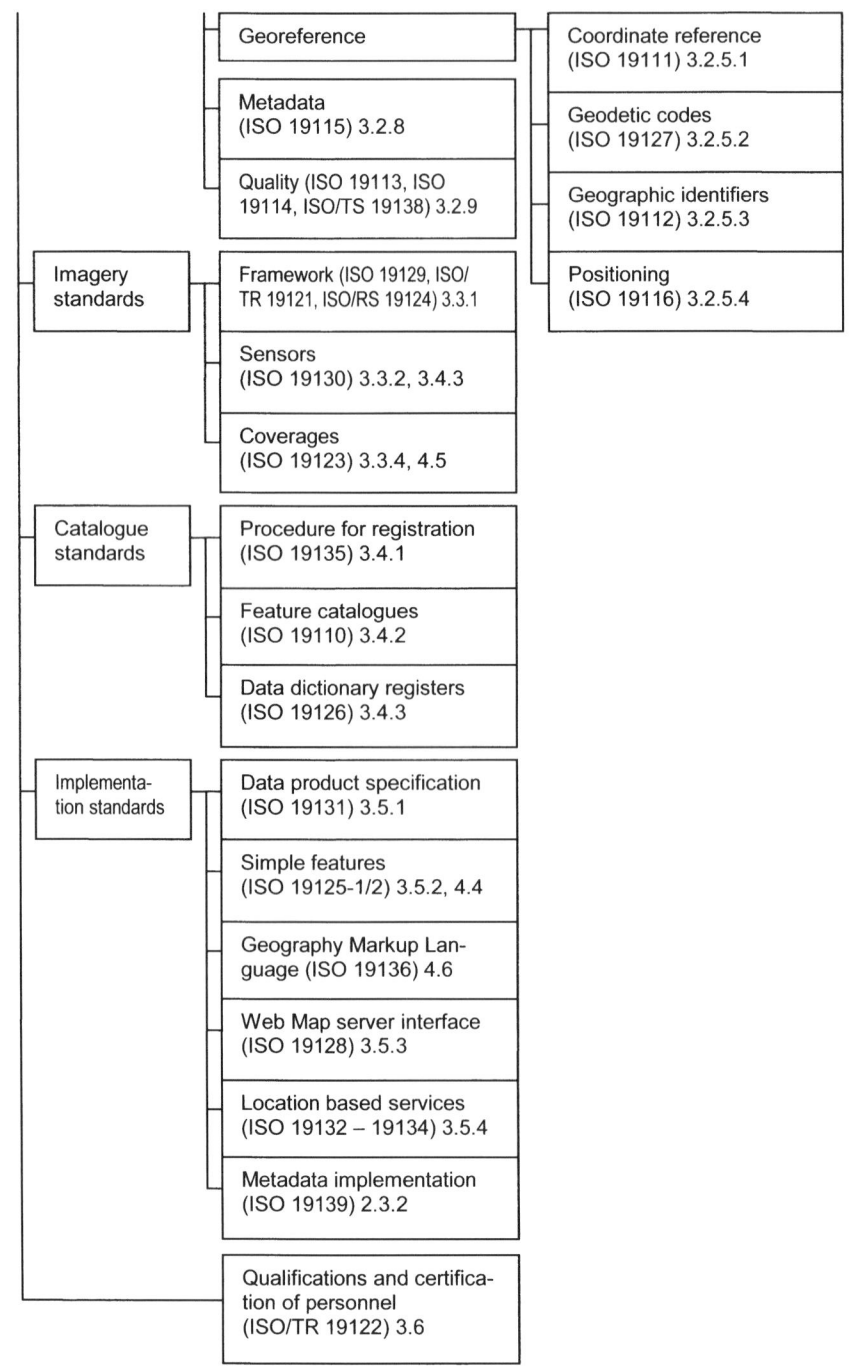

Fig. 2.6. Roadmap to the ISO 19100 standards

3 Non-geometry standards

Chapter 3 explains all ISO 19100 standards apart from ISO 19101 (Reference model) and the geometry-oriented standards that are ISO 19107 (Spatial schema), ISO 19123 (Schema for coverage geometry and functions), ISO 19125-1 (Simple feature access – Part 1: Common architecture), and ISO 19136 (Geography Markup Language). The ISO 19101 is explained in chapter 2. The other four standards are explained in detail in chapter 4.

3.1 Infrastructure standards

The infrastructure standards set rules that apply to all pieces of the ISO 19100 family. They define the "infrastructure" for the development of the standards themselves and for the development of application schemas and profiles. The infrastructure standards include the conceptual schema language (ISO/TS 19103), the terminology (ISO 19104), the conformance and testing (ISO 19105), and the profile (ISO 19106). Originally, the ISO 19102 (Overview) was meant to provide a general introduction to the ISO 19100 family. The project was later cancelled because it was difficult to continuously update it while standards evolved. The Internet and textbooks such as this, may provide a much better access to the ISO 19100 standards.

3.1.1 Conceptual schema language (ISO/TS 19103)

Today, if experts meet to discuss the design of a computer system they would talk in a conceptual schema language. Admittedly, they speak in English, French, or German, or any other language of the world, but a conversation like this remains informal and fuzzy until someone starts drawing a diagram, mostly in UML, the Unified Modelling Language, in order to express the ideas with the formal tools of classes or packages and their relationships. Therefore, it is essential for everybody who works in this field to be able to communicate in a conceptual schema language. The ISO/TS 19103 (Conceptual schema language) defines a UML profile for geographic information. This book assumes the reader has a basic knowledge of UML and focuses on the extensions of UML for geographic information.

Background

A conceptual schema language is based upon a conceptual formalism that provides the rules, constraints, inheritance mechanisms, events, functions, processes and other elements that make up a conceptual schema language. For the ISO 19100 family of standards the applicable conceptual formalism is the object-oriented modelling as described by OMG (OMG 2003). A conceptual schema language has to be capable of representing 100% of the semantics in a domain of discourse (see section 2.2.3). The 100%-requirement refers to the level of detail that is appropriate for modelling the domain in question. Traditional conceptual schemata such as the Entity-Relation model cannot describe numerical or logical relationships between values of concept. Therefore they are not able to meet the 100% requirement.

The UML has become the strongest of several conceptual schema languages that have been developed over the last decade. The roots of UML were independent but similar to developments by three well-known American "software methodologists": Booch, Rumbaugh, and Jacobson. These three later pooled their efforts and created a company, Rational Software Corporation, that has become the leading developer of software engineering tools. UML is about to become an International Standard prepared by the ISO/IEC JTC1/SC7 (ISOIEC47 2003). Today Rational Software Corporation is a division of IBM.

EXPRESS is a conceptual schema language being used in the field of mechanical engineering and was standardised by the ISO/TC184 (Industrial automation systems and integration) (ISOIEC25 1994). Conceptual schemas in UML are based on graphical and lexical elements, whereas the schemas of EXPRESS primarily rely on text. According to the standards of the ISO 19100 series, both languages are available for the modelling of geographic information. UML is preferred however, as it has turned out to be far more feasible for modelling geomatics. Therefore, this book uses UML as the only conceptual schema language.

UML elements for geographic information

The ISO/TS 19103 (Conceptual schema language) requires the use of UML as it is defined in the ISO/IEC 19501-1. Specific rules and recommendations have been established for the following aspects: classes, attributes, datatypes, operations, associations, and stereotypes. In addition, naming conventions and modelling guidelines maintain the unique appearance of the whole family of ISO 19100 standards (see annex E).

Classes

Normative models use class-diagrams and package diagrams. Other UML diagram-types, such as use-case diagrams, may be used for information. All normative models contain complete definitions of attributes, associations, operations, and appropriate data type definitions.

According to the ISO 19100 family, a class is viewed as a specification and not as an implementation. Attributes are considered to be abstract, and do not have to be directly implemented. For each class defined according to the ISO 19100 family, its set of defined attributes together with the sets of attributes of other classes (that are accessible either directly or indirectly via associations), shall be sufficient to fully support the implementation of each operation defined for this particular class.

Attributes

All attributes must be typed, and the type must exist among the set of legal base types. A type must always be specified; there is no default type.

Data types

Tables 3.1 to 3.3 show all the data types defined by the ISO/TS 19103. Some of the data types are adopted from UML while others are adopted from the Object Constraint Language (OCL) that is a component of the comprehensive UML standard. Some other data types belong to both.

The ISO/TS 19103 groups the basic types into three categories:

- Primitive types,
- Implementation types, and
- Derived types

In several operations NULL and EMPTY are possible values.

The Primitive types are the fundamental types for representing values.

Table 3.1. Primitive types in ISO/TS 19103

Data type	Examples	Defined in	
		UML	OCL
Integer	123, -65547	x	x
Decimal	12.34		
Real	12.34, -1.234E-4	x	x
Vector	(123, 456, 789)		
CharacterString	"This is a nice place"	x	x
Date	2003-02-19		
Time	13.59:30 or 13:59:30-05:00	x	
DateTime	2003-02-19T13:59:30		
Boolean	TRUE, FALSE	x	x
Logical	TRUE, FALSE, MAYBE		
Probability	$0.0 \leq p \leq 1.0$		
Multiplicity	1..*	x	

The Implementation types are template types for representing multiple occurrences of other types.

Table 3.2. Implementation types in ISO/TS 19103

Implementation type		Defined in OCL
Collection types	**Explanation**	
Set	collection, each object appears only once	x
Bag	collection, each object may appear more than once	x
Sequence	bag-like structure that orders the element instances	x
Dictionary	array of elements with an integer-index	
Enumeration types	**Examples**	
Enumeration	{Public, Private, Tourist}	x
CodeList	{2001, 2002, 2003}	
Representation types	**Examples**	
Record, Record type	(Ottawa, 800.000), (Neubrandenburg, 70.000)	
Generic name	GM_Object, TP_Object	

The Derived types are the measure types and the Units of measurement (Uom).

Table 3.3. Derived types in ISO/TS 19103

Measure type	Unit of measurement
Area	UomArea
Length, Distance	UomLength
Angle	UomAngle
Scale	UomScale
Mtime (measured time)	UomTime
Volume	UomVolume
Velocity	UomVelocity

Operations

An operation specifies a transformation on the state of the target object (and possibly the state of the rest of the system reachable from the target object), or a query that returns a value to the caller of the operation (Rumbaugh et al. 1999).

In UML, objects are normally modified or accessed by their own methods. However, objects can be "passed by reference" so that any persistent object passed as a parameter may receive a cascading message. The object can therefore be modified, albeit indirectly, by any method to which it is passed. Thus object-valued parameters are inherently "inout" in direction. Data types valued are "in" in the parameter list and "return" in the return list.

The current UML syntax allows for only one return value. If there are multiple return values they can be put in the parameter list with an "out" direction. Modified persistent objects that appear in the parameter list are implied to be "inout". All other parameters are "in".

Relationships and associations

A relationship in UML is a ratified semantic connection among model elements. Generalization, dependency, and refinement are class-to-class relationships. In the 19100 family of standards, they are used according to the standard UML notation and usage.

Association, aggregation, and composition are object-to-object relationships. An association is used to describe a relationship between two or more classes. An aggregation is a relationship between two classes, in which one of the classes plays the role of a container and the other plays the role of a content. A composition is a strong aggregation. In a composition, if a container object is deleted, then all of its content objects are deleted as well.

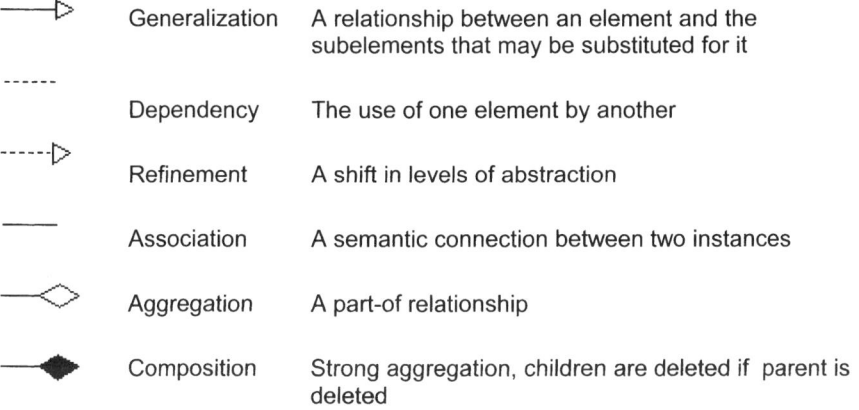

Fig. 3.1. Kinds of relationships in UML (ISO35 2001)

Stereotypes

Stereotypes are a method of classifying the UML-classes in order to augment the readability of larger UML class- and package-diagrams. Stereotypes indicate the context in which a class shall be applied. The ISO/TS 19130 defines 11 stereotypes as being relevant for geographic information. Nine of them are standard UML stereotypes. Stereotypes are one of the three extensibility mechanisms in UML. The other two are constraints and tagged values.

60   3 Non-geometry standards

- \<\<Interface\>\>
- \<\<Type\>\>
- \<\<Control\>\>
- \<\<Entity\>\>
- \<\<Boundary\>\>
- \<\<Enumeration\>\>
- \<\<Exception\>\>
- \<\<MetaClass\>\>
- \<\<DataType\>\>

The ISO/TS 19103 introduces two new stereotypes for use in geomatics.

\<\<CodeList\>\> is a list of potential and known values. An example is the list of curve interpolations in ISO 19107 (Spatial schema) {linear, geodesic, circular Arc3Points, etc.}. A \<\<CodeList\>\> is similar to an enumeration in that one of a number of values is possible, but each differs in intent as a code list may be expanded over time.

\<\<Leaf\>\> is a package that contains no sub-packages, only object classes and interface definitions.

Example for an UML class-diagram of a small GIS

The purpose of this example is to clarify the use of UML class-diagrams in the context of geographic information.

The purpose of the example-GIS is the computation of the fastest route between two cities (Figure 3.2).

The sketch-map shows the cities and two roads connecting them, including the required attributes about road condition and speed limits. Other aspects, like the population of the cities, are outside the model because they are irrelevant to the purpose of the GIS.

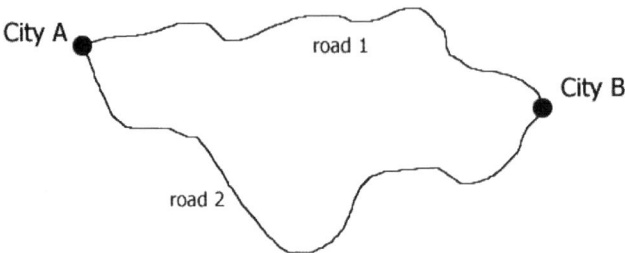

Fig. 3.2.: Example-GIS with two roads and two cities

3.1 Infrastructure standards

The application program used for the route computation requires the data of the GIS. The data are partitioned into the dataset and the metadata.

The program involved falls in the category of geographic information services. This service is a software program that uses the data supplied in the dataset and in the metadata.

The dataset contains the cities and the roads, and the underlying geometries such as points, lines, and coordinates. The cities and roads are called geographic features. A specific city A and another city B are called feature instances. The points and lines are called spatial objects and the coordinates are referred to as DirectPositions.

The metadata contains attributes like road condition and speed limit. In this example one metadata exists for one feature instance of a road.

The most frequently used type of UML-diagram is a class-diagram which consists of classes and their interrelations. A class may be anything that the developer has defined to have common properties. In this example the first case three classes have been defined:

- City
- Road
- RouteFinder

The class RouteFinder may be called a title class of the classes Road and City. The class RouteFinder can also be called parent-class whereas the other two are the children-classes.

Later in the development the model might have to be refined. Let us assume that different types of computations are required for routes using the main roads and the local roads. For this purpose, the class Roads would be split into subclasses according to the road classification. As far as possible, properties like the length of the road would be kept in the super class as feasible. Only the properties that are unique to a subclass such as the maximum speed or winter closure have to be placed there.

A UML class has a name, a set of attributes, a set of operations, and constraints. In our example, the class names are City and Road. It is required that a class name is unique in a model. Thus, in larger models the class name is often a combination of the identification of the sub-model and the name like FR_City and FR_Road, where FR stands for the program's purpose "fastest route". Class names must start with an uppercase character.

The second case of the model that includes the main road and the local road is shown in figure 3.3.

Attributes are values that are related to the class. In our example, the length of the road is an attribute. Another attribute may be a status tag like "road open" or "road closed".

Operations are functions that are related to the class. Many classes have the two rather basic functions "setAttributes" and "getAttributes" that allow writing and reading of the attributes of the class. Another operation that might make some sense in our example could answer the question of whether the whole length of the road can be travelled within one hour going at a speed of 80 km/h. The function would address the attributes length and availability, and return the result.

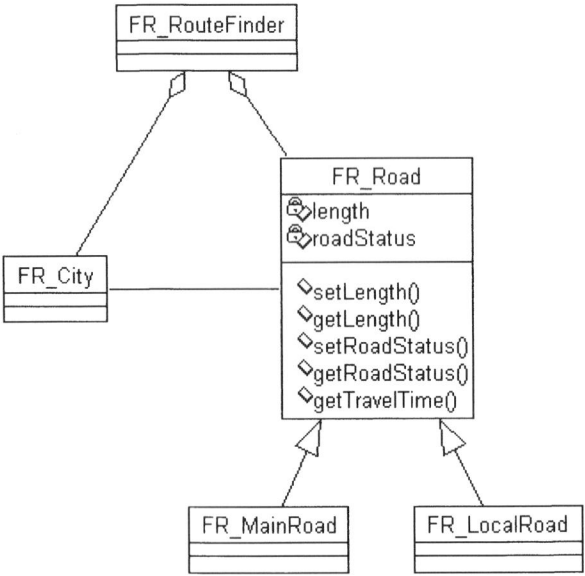

Fig. 3.3. UML class-diagram of the route-finder example (FR = fastest route)

Constraints are limitations within a class or within other parts of the model. In our example the number of values per attribute are constrained to be one value. This reflects the fact that a road can usually have only one length. But we can easily think of other road-attributes that may exist many times, such as intersections with other roads.

Associations are semantic connections between classes. It is easy to understand that the class RouteFinder is associated with the classes Road and City as both are subclasses of the class RouteFinder. The class Road is also directly associated to the class City because roads start and end at cities. Finally, the class Road has associations to its subclasses MainRoad and LocalRoad.

The quality of these associations differs slightly. The route-finder system is composed of cities and roads. This part of the relationship is called an aggregation. Both, Road and City, have a relationship called aggregation toward the class RouteFinder. The class RouteFinder is called the parent-class and the classes Road and City are called the children.

If we consider our route-finder system as being complete without any plans of later expansion, the classes Road and City are essential for the life of the system. A strong dependency exists between the children and the parent-class and without these classes, the system would no longer work. For example, if the class Road should be deleted, the class RouteFinder no longer makes sense and must be deleted as well. In this case, the association between both classes is strong and is called a composition.

If we consider our model being open towards other applications in the future we can keep the term aggregation. An example of a future application might be the search for the fastest railway connection. A new class Railway would have to be created and then aggregated to the class RouteFinder. In this scenario, we should keep the association between the classes RouteFinder and Roads on a weaker level and set it to aggregation only. With an aggregation we can indicate that our route finder system is not just bound to the first application in place.

The classes MainRoad and LocalRoad are specialisations of the class Road. A road is either a main road or a local road but UML also puts it the other way round: the association between the class Road and the classes MainRoad and LocalRoad is called generalization.

UML class-diagrams are used in all stages of a development process. An UML-model may contain the final product with all classes well defined. It may also display the system in an immature state with a number of open questions within the model. Let us imagine that the development of our route-finder GIS started before it has been decided whether it will calculate routes on roads or on railways or on both. In this case it is helpful to create a so-called abstract class TransportationLines without determining ahead of time whether it will represent roads, railways, or both.

The following three examples show the application of three standards of the ISO 19100 family.

Example for UML class-diagrams taken from ISO 19100 standards

The example has been taken from the ISO 19111 (Spatial referencing by coordinates). Basically, it consists of two generalization trees, one with the root class RS_ReferenceSystem and the other with the root class SC_Datum. The trees are associated with a number of other classes as well as with themselves. Enumerations define the attribute values. Notes are used for clarification. The class-diagram is normative.

64   3 Non-geometry standards

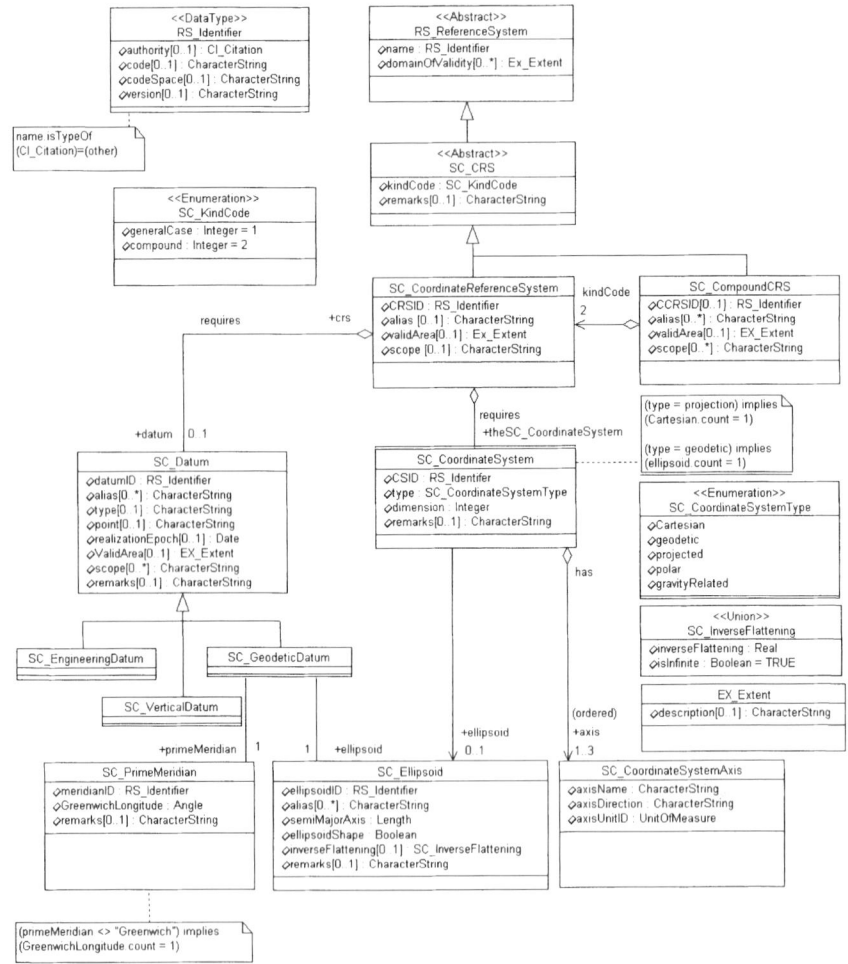

Fig. 3.4. UML class-diagram from ISO 19111 (Spatial referencing by coordinates) (ISO43 2003)

Example for a UML package-diagram

The example has been taken from the ISO 19107 (Spatial schema). It displays the geometry package with the class content and the internal dependencies. The package-diagram is normative.

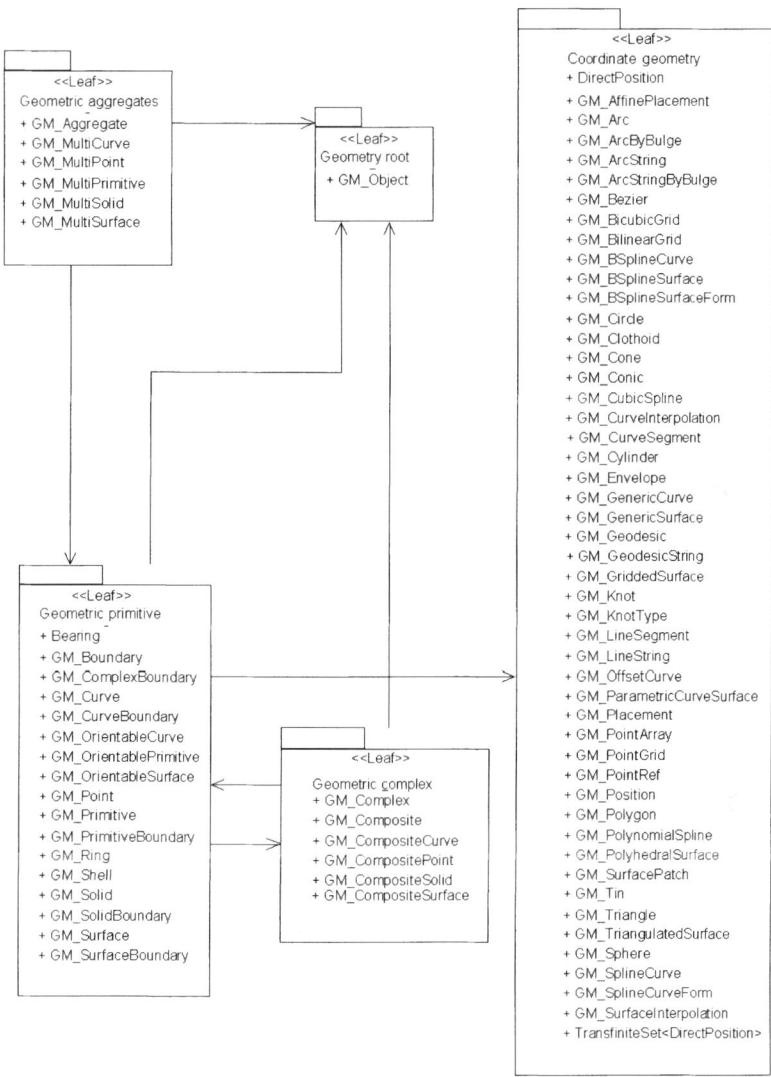

Fig. 3.5. UML package-diagram from ISO 19107 (Spatial schema, geometry package) (ISO39 2003)

Example for a UML use-case diagram

The example has been taken from the ISO 19116 (Positioning services). It illustrates the use-cases for which the standard is primarily intended. The use-case diagram is informative and, therefore, is not normative.

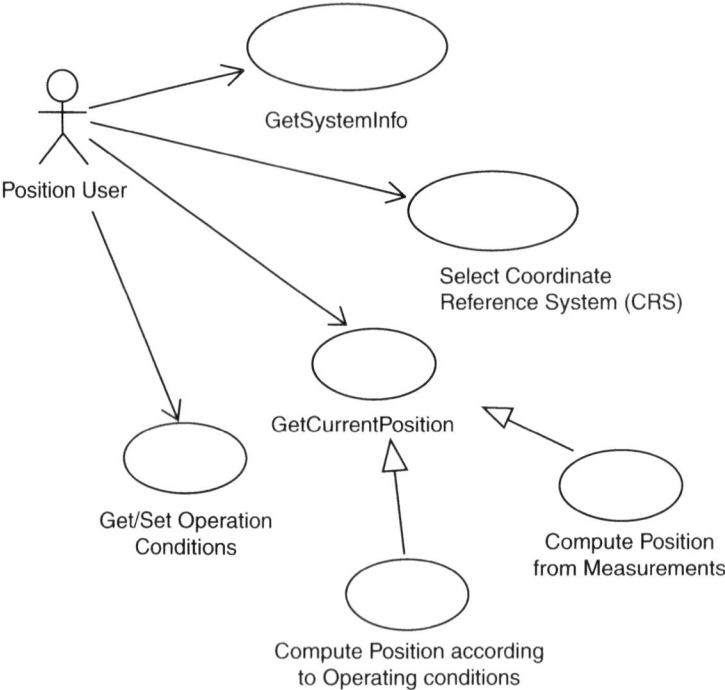

Fig. 3.6. UML use-case diagram from ISO 19116 (Positioning services) (ISO48 2002)

3.1.2 Terminology (ISO 19104)

The ISO 19104 (Terminology) provides guidelines for collection and maintenance of terminology in the field of geomatics. The main concept for setting up terminology demands that the same term is used for the same concept throughout the whole family of standards. The ISO 19104 also lays down the guidelines for maintenance of the terminology repository.

Each standard addresses its segment of geographic information and uses the most appropriate terms found there. The ISO standard "Terminology" guarantees the consistency of all involved terms through the specification of a terminological record and the description of principles for definition writing.

In practice, a consistent terminology among all ISO 19100 standards has not yet been achieved. This is due to the fact that some of the standards have been developed outside the ISO/TC211 environment. Examples are the ISO 19123 (Schema for coverage geometry and functions) that has been proposed by the Open GIS Consortium, and the ISO 19136 (Geography Markup Language) that has been originally developed by a private company.

Much of the terminology has now been harmonized. The terms that still have to be kept with different meanings are marked in the list of definitions.

As ISO standards have to be reviewed every five years, a family of standards undergoes minor, but continuous, change. In order to guarantee a current and efficient management of the common terms, a terminology repository is planned. A preliminary version is available as an online database that can be accessed publicly via the Internet (SIA 2003).

The database covers all ISO 19100 terms, showing their definitions, and providing notice where multiple concepts for the same term apply. The database is continuously updated.

3.1.3 Conformance and testing (ISO 19105)

If it is claimed that a data file is written in a standardised format like TIFF, then it would be easy to test the conformance with the standard. If a correct TIFF-reader is able to open and display the file then, as far as the content is concerned, both the file and the reader conform to the standard. In larger systems however, it becomes more difficult to decide if they conform to certain standards. In order to execute a test that is independent of the system manufacturer and the user, independent institutions such as testing laboratories and accreditation bodies, often become involved. Testing of conformance represents a major step during the introduction of the system to the market. The ISO 19105 (Conformance and testing) sets the rules for the conformance tests of all ISO-standards for geographic information.

Similar to many other basics of the ISO-standards for geographic information, the original rules for conformance testing were developed by the Information Technology community (ISOIEC17 1994; ISOIEC18 1994; ISOIEC19 1994). These rules were enhanced for use within the ISO 19100 series. Conformance testing means the testing of a candidate product for the existence of specific characteristics. The testing addresses the capabilities of an implementation compared to the:

1. Conformance requirements as defined in each standard document.
2. Product description of the manufacturer.

The first is a test according to conformance class A and the second is a test according to conformance class B.

The ISO 19105 defines two types of conformance tests. The "basic test" provides limited testing of an implementation under test (IUT) in order to establish whether or not it is appropriate to perform more thorough testing. The "capability test" should exercise an implementation as thoroughly as is practical over the full range of conformance requirements specified in the standard.

The conformance testing does not include testing the robustness of an implementation, the acceptance at the client's site, or the system's performance.

All testable ISO geographic information standards contain a conformance clause that specifies all the requirements that must be satisfied in order to claim conformance to that standard. The conformance clause serves as an entry point for conformance testing. The conformance clause is hierarchically structured into the upper level of an abstract test suite, the medium level of abstract test modules, and the lower level of abstract test cases. The precise definition of the test purpose is the key statement of every test module and every test case. An example of a test purpose may be: "Test the generation of a polygonal line as a sequence without self intersection." This is a little example, but it indicates that the abstract test suite of the major standards may become rather voluminous documents. It is for the standard's developer to create a conformance clause that includes all test methods and test cases necessary to guarantee a complete conformance to the standard.

An implementation that is to undergo a conformance test is called an "Implementation Under Test" (IUT). The ISO 19105 (Conformance and testing) structures the test in four steps: preparation for testing, test campaign, analysis of results, and conformance test report. The conformance assessment process is carried out by an independent testing laboratory.

The formalised approach to the testing of implementations has some important intrinsic properties. A test must be repeatable in that two or more tests of the same implementation are comparable and have mainly the same results. The test is also auditable in that the work of the independent testing laboratory may be subject to audit.

3.1.4 Profiles

3.1.4.1 Creation of profiles (ISO 19106)

The family of International Standards for geographic information covers an immense range of possible applications with reference to the earth. But the complete ISO 19100 family of regulations would overload most real applications. Therefore, only a subset of the ISO 19100 standards is normally required. A profile according to ISO 19106 (Profiles) is a subset of the ISO 19100 standards.

The ISO 19100 standards provide two approaches for the creation of a specific application, profiles, and application schemas. While a profile narrows the functionality, an application schema extends it beyond the scope of a given standard in order to meet specific needs. Accordingly an application schema is developed in two steps:

1. Definition of a profile of the ISO 19100 standards
2. Creation of an application according to ISO 19101 (Reference model) and ISO 19109 (Rules for application schema) for all additional components.

3.1 Infrastructure standards 69

A profile may become a standard on its own and is then called an International Standardised Profile (ISP). In the mid-nineties, the Joint Technical Committee of ISO and IEC (ISO/IEC JTC1) created the International Standardised Profile as a new type of document. In order to receive the status of an ISO standard, the document describing a profile has to follow the procedures for the development of an International Standard. As a result, it receives its own ISO-number. Profiles defined according to the ISO standard "Profile" may become this type of an International Standard.

One may argue that in times of fast computers and the availability of enormous amounts of disc space, it is not necessary to artificially narrow the options offered by a large environment like the ISO 19100 family of standards. However, the idea of the introduction of profiles is to promote a better interoperability between the systems by restricting the choices. It is easier for users and system suppliers to agree on a smaller set of common standards then on a large set.

C. Douglas O'Brien has expressed interest in the area of spatial data engineering from his university days where his thesis topic was the development of the computer programming language 'IMAGE Interactive Manipulation of A Graphics Environment'. Today Mr. O'Brien is president of IDON Technologies Inc.

Mr. O'Brien was born in Ottawa, Canada, and received his Master's degree from Carleton University. He is the inventor of the PDIs (Picture Description Instructions) which formed the basis of the North America videotex and teletex standards.

Later he became the Director of Systems Technology Research and Development of the Canadian Department of Communications. He first became involved in the area of standardisation with respect to Computer Graphic standards through the Association of Computing Machinery in 1975.

The ISO/TC211 activities have been shaped by Mr. O'Brien from the very beginning – Canada proposed the Technical Committee, Mr. O'Brien wrote the original proposal. Later he became project leader of ISO 19106 (Profiles) and convenor of the Working Groups 5 (Profiles and functional standards) followed by Working Group 6 (Imagery). Mr. O'Brien is one of the members of ISO/TC211 with the most comprehensive knowledge on standardisation procedures and background information.

He is currently involved in many other aspects of geomatics including the NATO Digital Geographic Information Working Group, the International Hydrographic Organization, and the ISO/IEC JTC1 SC24 (Computer graphics and image processing).

70   3 Non-geometry standards

How is a profile defined?
- A profile is a subset of the base standards of the ISO 19100 family of standards or other Information Technology standards.
- A profile determines how they are used together.
- A profile explains the usage details as far as required.

ISO 19106 distinguishes between two types of profiles.
- The conformance level 1 designates a profile that is purely derived from elements of the ISO 19100 family of standards and possibly other ISO standards.
- The conformance level 2 specifies profiles that integrate elements of non-ISO standards with elements from the ISO standards.

The development of profiles has been opened according to conformance level 2 because the existing suite of standards cannot yet claim to meet *all* requirements. This is despite the great effort that has been put into a generalized and comprehensive approach towards geographic information. A profile of conformance level 2 cannot become an ISO standard of the type International Standardised Profile.

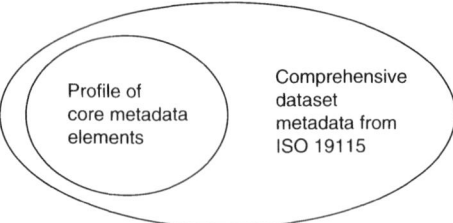

Fig. 3.7. Example of a profile using concepts and structures from one standard (ISO38 2002)

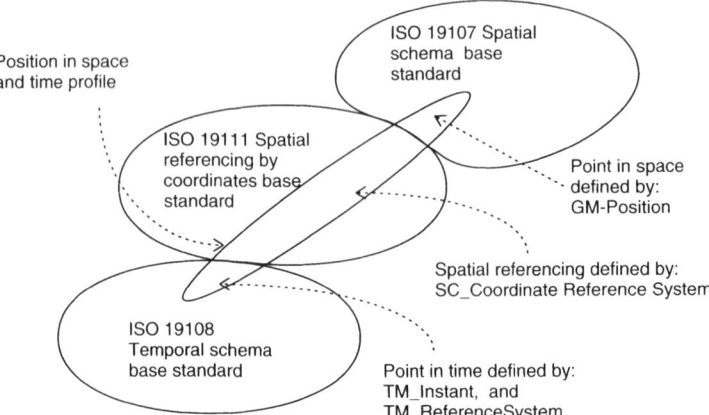

Fig. 3.8. Example of a profile using concepts and structures from more than one standard (ISO38 2002)

3.1.4.2 Functional standards (ISO/TR 19120)

From a long-term perspective, the standards of the ISO 19100 family will become the only internationally agreed foundation for geographic Information Technology. However, while ISO has been developing the standards, other international organizations have already agreed on their geographic data environments. Thus, it is ISO's objective to develop a generic suite of standards that is capable of integrating existing environments. It is the challenge of ISO 19120 (Functional standards) to identify the existing environments and to provide assistance with the development of profiles.

A functional standard is an existing geographic information data standard developed specifically for the transfer of data between entities in different nations. The efforts of ISO/TR 19120 resulted in a report that considers three functional standards that should be harmonised with the ISO 19100 series.

- The Digital Geographic Exchange Standard (DIGEST) is used to support the military digital geographic information requirements amongst NATO nations. The standard is maintained by the Digital Geographic Information Working Group (DGIWG).
- The Geographic Data Files (GDF) is used to define and exchange digital road databases with a particular emphasis on navigation applications. The ISO/TC204 (Intelligent transport systems) has created an International Standard for global GDF.
- The International Hydrographic Organization (IHO) Transfer Standard S-57 is intended to be used for the exchange of digital hydrographic data between national hydrographic offices and for distribution to manufacturers, mariners, and other data users.

As the ISO 19100 standards become available, other functional standards are likely to be identified in the future. Therefore, the report is only considered the starting point for a feedback cycle between the functional standards communities and the ISO/TC211 experts and its function is to support the maintenance and future revisions of the ISO 19100 standards. It is intended to identify components and elements that can be harmonised between both communities. This feedback is made available to ISO/TC211 through amendments to the initial report.

3.2 Basic standards

As stated in chapter 2, section 4 the members of the ISO 19100 family of standards can be unofficially categorized into the groups infrastructure, basic, imagery, catalogue, and implementation standards. The basic standards comprise the aspects of services, space, time, georeference, portrayal, encoding, metadata, quality, and application schema.

3.2.1 Services (ISO 19119)

The Information Technology community decomposes computer software systems by using five different viewpoints. One of them, the computational viewpoint, addresses the services. The ISO 19119 (Services) provides the framework to structure services in the context of geographic information. In the case of geomatics, services may be pure Information Technology services like querying a database or specific geographic information services such as finding a location in a Coordinate Reference System.

Services are defined as the capability to provide for manipulating, transforming, managing, or presenting information. A special case is the service interfaces that are the boundaries across which services are invoked and across which data is passed between a service and an application, external storage device, communications network, or a human being. Following these definitions, the ISO 19100 standards fall into two categories: infrastructure standards and service standards. Infrastructure standards are guidelines that are applicable for all other standards. Examples are the ISO 19101 (Reference model) and ISO 19104 (Terminology). With few exceptions, all standards with numbers equal or higher than ISO 19107 are service standards.

The ISO 19101 (Reference model) sets forth the conceptual modelling, the domain reference model and the architectural reference model. The last embraces the IT-services as defined in the Reference Model of Open Distributed Processing (RM-ODP) and standardises the Extended Open System Environment Reference Model (EOSE-RM) for the specific services for geographic information. The general Open System Environment Reference Model (OSE-RM) is standardised in the ISO/IEC TR 14252.

Figure 3.9 summarises all IT- and geographic information services. Without claiming completeness, the diagram also relates the ISO 19100 standards to the three service categories of geographic information. The acronyms are explained below.

3.2 Basic standards 73

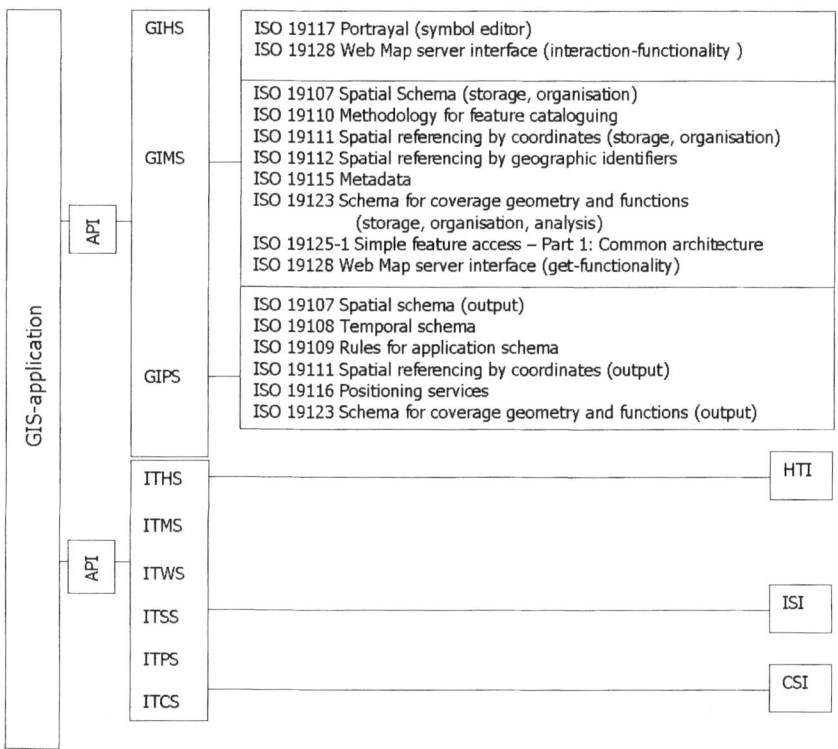

Fig. 3.9.: Architectural reference model showing the services and their relations (ISO34 2002)

The OSE-RM defines six types of services and five types of interfaces through which the services communicate with other applications or peripheral devices.

IT Service types:
1. Human Interaction Service (ITHS). An example is the interaction of an operator with the graphic interface while logging on.

2. Model Management Service (ITMS). An example is the software-supported creation of an UML-diagram while designing a geographic application.

3. Workflow/Task Service (ITWS). An example is the setting of a later start time for an overnight run of a large processing job.

4. System Management Service (ITSS). An example is the creation of a new user with access rights to a geographic database.

5. Processing Service (ITPS). An example is the calculation of the average density of population in a given dataset.

6. Communication Service (ITCS). An example is the provision of map data over the Internet.

Interface types:
1. Application Programming Interface (API). All services communicate via the API with the GIS applications.

2. Human Technology Interface (HTI). The best-known examples are the window systems on the screen.

3. Information Service Interface (ISI). The ISI establishes the link to all data sources like the databases where data shall be retrieved.

4. Communications Service Interface (CSI). The CSI is the link between the computer and the network.

5. Network-to-Network interface. This is the network itself.

Antony Cooper is South African. He was born in Johannesburg and raised in Grahamstown (Eastern Cape), at the centre of where the 1820 Settlers from Britain were settled. He is a Divisional Fellow in the Information and Communications Technology Group of the CSIR a statutory research council in Pretoria. Within ISO/TC211 he has convened Working Groups 2 and 7 and acted as the Head of Delegation for South Africa at many plenary meetings.

Mr. Cooper hold a BSc, a BSc (Honours) and a MSc in Computer Science. The majors range from Cost and Management Accounting to the exchange of digital geographic data. He has applied his expertise to a wide range of applications, including crime mapping, siting polling stations, quality assurance, and integrated development planning. In 1999 he and a colleague won the first prize for the most innovative use of mapping at the first International Crime Mapping Research Competition in Orlando, Florida.

Mr. Cooper's interests include playing, umpiring and coaching field hockey, playing cricket, bird watching and genealogical research.

In addition, the EOSE-RM includes the following services. Theoretically every IT service has an equivalent geographic information service. In fact, only three categories are relevant for geographic information:

GI Service types:
1. Geographic Information Human Interaction Service (GIHS). An example is the interaction of an operator with the graphic interface while picking a line.

2. Geographic Information Model Management Service (GIMS). An example is the storage of a geographic feature according to an application schema.

3. Geographic Information Processing Service (GIPS). An example is the output of a map to a screen.

3.2.2 Rules for application schemas (ISO 19109)

The ISO 19100 standards address the full range of geographic information. A certain application only uses a subset of the available standards combined with many additional details that are beyond the scope of the abstract standards. An example is a feature catalogue for a given application.

The ISO 19100 standards provide two approaches to creating a specific application: profiles and application schemas. See chapter 3, section 1.4.1 "Profiles" for further details.

The kernel of ISO 19109 is the definition of a geographic feature. A feature stands for anything in the real world. For example, a feature can be a single corner stone of a land parcel, a whole country, a digital elevation model, or a satellite image. In order to integrate a feature into the models of geographic information in a homogeneous way, the ISO 19109 defines the "General feature model" (GFM).

The GFM defines an abstract feature with attributes and operations. Attributes contain all the static information such as the quality of the feature or its geometric properties (point, curve, surface, or solid). Operations contain information about the change of a feature according to external influences like a road being closed in winter or a road being displayed only within the scale range 1:5,000 to 1:25,000. This change is also called the behaviour of the feature.

As the examples show, features can differ in importance and size. In practice, this often leads to a hierarchical grouping of features. For example, public buildings and private buildings are both "buildings". The GFM allows for the construction of a generalization tree where the feature types (class GF_FeatureType) public and private buildings are specialisations within the feature type building.

Generally features reside independently in the dataset. In the case of a 3-dimensional dataset the relations between features can be computed in all spatial dimensions. The over-, under-, or level-situation at intersections is a particularly important case for mapping and network computation. A 3-dimensional dataset implicitly contains the necessary information. A 2-dimensional dataset does not allow for these computations. In order to supply the missing information, the GFM includes associations between features that contain the information regarding which feature is above the other or if both are on the same level. In this case, the association type (class GF_AssociationType) would be "intersection", and the association roles (class GF_AssociationRole) would be under, over, or level respectively.

In some cases, it may be advisable to impose constraints to the definition of a feature. A theoretical example may be that a feature of type curve must not have more than 1000 points. The GFM allows the formulating of these constraints (class GF_Constraint).

An application schema is usually created by the definition of the features. All details are collected in a feature catalogue. The methodology for building feature catalogues is covered by ISO 19110.

3.2.3 Spatial schema (ISO 19107)

ISO 19107 (Spatial schema) is a standard to describe the geometry and the topology of geographic information. This standard comprises a comprehensive definition of the geometric elements required to build a geographic dataset.

The standard primarily addresses vector data up to three dimensions. For the 2-dimensional case, the standard includes provisions that guarantee a seamless coverage of a complete area.

The ISO 10107 defines a method to describe the position of a geometric element. A position is named DirectPosition. It includes the coordinates such as x, y, and z depending on the Coordinate Reference System of the dataset and its dimension. This method is used throughout other important standards of the ISO 19100 family such as the ISO 19123 (Schema for coverage geometry and functions).

This standard does not address the graphic portrayal of the geometric elements.

The chapter 4, section 3 "Spatial schema" contains a detailed description of ISO 19107.

3.2.4 Temporal schema (ISO 19108)

As far as geomatics is concerned, temporal characteristics are standardised in ISO 19108 (Temporal schema). This standard is not an independent standard for the description of time, but rather an addition to the existing standards (ISO05 1992). Many geographic applications require a time stamp related to the physical reality and therefore, the standard deals with the valid time which is the time when the event occurred in the abstracted reality. The standard does not address the transaction time when the data becomes available in a database.

In many applications, the time is handled as a metadata element only. More advanced applications require the time as a further dimension. This enables the modelling of the behaviour of a feature as a function of time, like a satellite's position along its orbit.

The standard distinguishes between the geometry of time and the topology of time. The geometry of time has the four major classes:

- The instant: a certain time
- The period: time elapsed between two instances
- The order: sequence of instances
- The relative position: timely relation of earlier and later instances

The topology of time describes the timely connectivity between two or more occurrences. If two or more occurrences take place at one instant, that instant is called a node. More precisely, the period between the occurrences is smaller than the resolution of the time. If two or more occurrences take place simultaneously during the period, that period is called an edge.

More timely aspects are a part of the ISO 19136 (Geography Markup Language), portrayed in chapter 4, section 6.

3.2.5 Georeference

This section addresses the ISO 19111 (Spatial referencing by coordinates), the ISO/TS 19127 (Geodetic codes and parameters), the ISO 19112 (Spatial referencing by geographic identifiers), and the ISO 19116 (Positioning services).

The ISO 6709:1983 (Standard representation of latitude, longitude and altitude for geographic point locations) has been developed by the ISO/IEC JTC1. In the year 2002 the maintenance of the ISO 6709 was transferred to the ISO/TC211. A systematic review of this standards revealed some shortcomings for geographic information purposes. Most probably the ISO 6709 will be amended in the future.

3.2.5.1 Coordinate reference (ISO 19111)

The ISO 19111 (Spatial referencing by coordinates) models coordinate systems and coordinate transformations for the ISO geomatics standards. This model contains the horizontal and vertical coordinate references for geographic features. Coordinate Reference Systems may be 1-, 2-, or 3-dimensional and do not change over time.

Ellipsoidal, Cartesian, and projected coordinate system

A Coordinate Reference System is a coordinate system that has a reference to the earth by a so called datum. The ISO 19111 standardises the details in order to fully define a Coordinate Reference System. The coordinate system may be an ellipsoidal system, a Cartesian system, or a projected coordinate system. Projected coordinate systems are used to realise map projections.

If it is an ellipsoidal system, the three coordinates are longitude, latitude, and ellipsoidal height. In most applications longitude and latitude are referred to as geographic coordinates. Because ellipsoidal heights are sometimes difficult to relate to the topographic earth's surface, the third coordinate is usually not related to the ellipsoid. Instead it could be given in an independent elevation reference system. This method is acceptable, because in small-scale and medium-scale applications, the decrease of the height accuracy caused by introducing a second reference system is insignificant compared to the accuracy of the height values.

The origin of a Cartesian System lies close to the centre of mass of the earth with the equator-plane of the earth usually defining the xy-plane of the system. The positive x-axis points to the intersection of the Greenwich meridian with the equator – a point off the African Atlantic coast. The positive z-axis points to the North Pole and the xyz-system realises a right hand system. Cartesian systems have become popular since they are used for describing the orbits of navigation satellites such as GPS.

A projected coordinate system is usually a two-dimensional system referred to a geometric developable surface like a cylinder, a cone, or a plane. These surfaces realise a local approximation of the earth's surface. In most cases, one axis of the system points north or south, the other east or west.

Datum and coordinate systems

A geodetic datum gives the relationship of a coordinate system to the earth. It is geodetic, vertical, or engineering. In many cases, it requires an ellipsoid definition. A vertical datum gives the relationship of gravity-related heights to a surface known as the geoid. The geoid is a surface close to mean sea level. According to ISO 19111, a datum shall be engineering if it is neither geodetic nor vertical.

A prime meridian defines the origin from which longitude values are specified. Most geodetic datums use Greenwich as their prime meridian.

A Coordinate Reference System is called fully defined if it has a datum and a coordinate system. Historically, the datum was often the position of an observatory near the centre of a country where the position and the orientation of the coordinate system are defined in relation to the physical reality of the earth. The prime meridian defines the origin of the coordinate values of the first ellipsoidal axis.

A reference that is defined by two different Coordinate Reference Systems - one of these is an elevation system - is called a *compound* Coordinate Reference System. A compound Coordinate Reference System is usually used to provide an independent reference for horizontal and vertical coordinates.

The standard does not address any specific coordinate system such as UTM (Universal Transverse Mercator) or WGS84 (World Geodetic System 1984). Also, the standard excludes the details of vertical systems bound to the physical structure and the real geometric shape of the earth being known as the geoid.

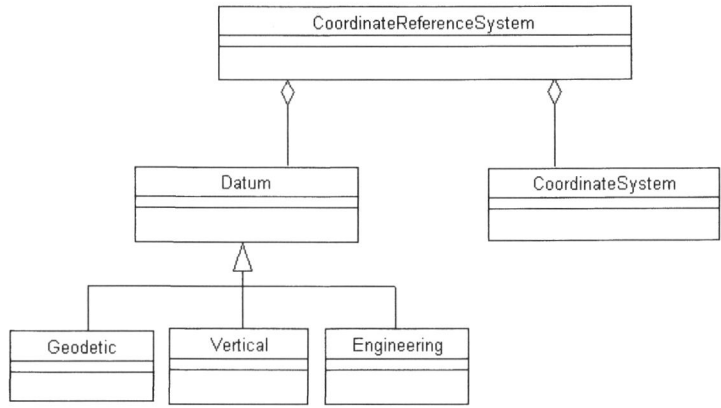

Fig. 3.10. Coordinate Reference System (ISO43 2003)

Conversion and transformation

The details of the specification of a Coordinate Reference System become important if datasets with a reference to two or more Coordinate Reference Systems are processed in the same environment. The coordinates of all features have to be made available in the same Coordinate Reference System. Consequently, positions given in other systems must be converted or transformed. The standard describes the information required to change coordinate values from one Coordinate Reference System to another by operations.

The operation is a generalized term for conversion and transformation (supertype). The standard also sets the frame for changing coordinates between two different Coordinate Reference Systems and distinguishes between "conversion" , where the involved Coordinate Reference Systems have the same datum, and "transformation", where the involved Coordinate Reference Systems have different datums.

A *conversion* changes coordinates from one coordinate system to another based on the same datum. In a coordinate conversion, the parameter values are exact. The coordinate conversion includes the map projection, the coordinate conversion of ellipsoidal coordinates to 3-dimensional Cartesian coordinates, unit changes, and the shifting of the origin towards a local grid. A map projection converts 3-dimensional ellipsoidal coordinates (excluding the height) to 2-dimensional Cartesian coordinates. In all cases, the conversion is based on exact formulas that are well known in the scientific literature and are quoted in the annex of the ISO 19111. An example for a map projection is the conversion between ellipsoidal coordinates on the Hayford ellipsoid to projected coordinates in UTM (Hayford ellipsoid), and vice versa. Elevations are not taken into account.

A *transformation* changes coordinates between different Coordinate Reference Systems with different datums. The shift, rotation, and a scaling between different Coordinate Reference Systems is derived from identical non error free defined

80 3 Non-geometry standards

points. The orientation of global Coordinate Reference Systems is always an estimation because it is based on a number of well defined (but differently selected) reference points where the residuals are minimised according to an adjustment method. Because the exact geometric relation between the Coordinate Reference Systems is not exactly known, a transformation can only be performed with an accuracy level that corresponds to the lowest accuracy of the definition of the system parts itself.

For example, if coordinates given in WGS84 (WGS84 ellipsoid) are needed in the German Gauss-Krüger System (Bessel ellipsoid), they must be transformed.

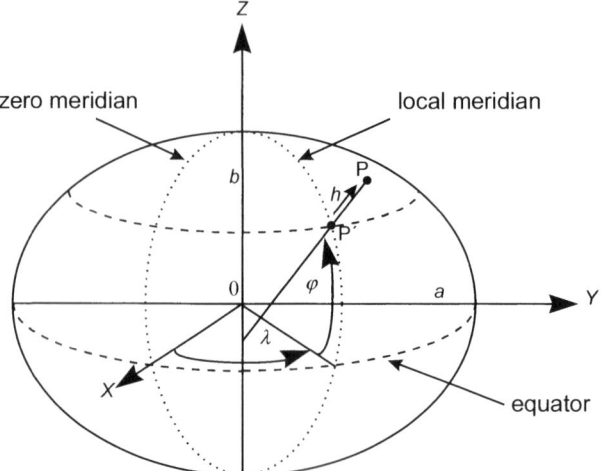

Fig. 3.11. Ellipsoidal and Cartesian coordinate system (ISO43 2003)

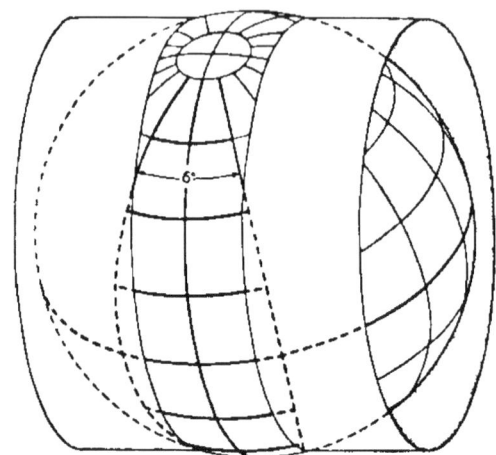

Fig. 3.12. Ellipsoidal and projected Coordinate Reference System (CRS) Example: Universal Transverse Mercator Projection (UTM) with strips 6° wide in longitude (Knickmeyer 2003)

In order to illustrate the meaning of the ISO 19111 the table 3.4. provides examples for different types of coordinates reference systems.

Table 3.4. Examples for different types of Coordinate Reference Systems (CRS)

Ellipsoidal CRS (longitude, latitude, ellipsoidal height)
WGS84 (φ, λ, h)   World Geodetic System 84 (1984)

Cartesian CRS (X, Y, Z)
WGS84 (X, Y, Z)   World Geodetic System 84 (1984)
ETRS 89 (X, Y, Z)   European Terrestrial Reference System 89 (1989)

Projected CRS (X, Y)
NAD83 / Alabama East   North American Datum 1983

Compound CRS: Ellipsoidal CRS + geoidal height (long., lat., gravity related height)
OSGB36 + ODN   Ordnance Survey of Great Britain 1936 +
                                 Ordnance Datum Newlyn

Compound CRS: Projected CRS + geoidal height (X, Y, gravity related height)
ETRS89 (projected) + EVRS2000   European Terrestrial Reference System 89 (1989) +
                                 European Vertical Reference System 2000

DHDN + DHHN   Deutsches Hauptdreiecksnetz (Gauss-Krüger
                                 Germany)
                                 Deutsches Haupthöhen Netz (German vertical datum)

3.2.5.2 Geodetic codes (ISO/TS 19127)

Within the ISO 19100 family of standards, the Coordinate Reference Systems are addressed in the ISO 19111 (Spatial referencing by coordinates) but they do not standardise a specific Coordinate Reference System. Therefore, the ISO Technical Specification ISO/TS 19127 (Geodetic codes and parameters) bridges the gap between the abstract frame of ISO 19111 and practical needs.

Within the geographic information community many references exist defining geodetic codes and parameters; none of which are in full compliance with the ISO 19111. The ISO/TS 19127 (Geodetic codes and parameters) provides the required guidance to apply the ISO 19111 (Spatial referencing by coordinates) in an appropriate manner.

The mechanism in the ISO 19100 family for creating publicly available lists of codes and parameters is a registry (ISO 19135). The ISO/TS 19127 provides rules for the creation and the maintenance of registers for geodetic codes and parameters.

In order to promote the use of the ISO 19100 model for coordinate reference, the ISO/TS 19127 contains a short list of Coordinate Reference System data and coordinate transformation data that are international in scope, widely used, well-defined, and unique. These data include the International Terrestrial Reference System

(ITRS), the Earth Gravity Model (EGM) 96 geopotential model, and the Universal Transverse Mercator (UTM) Coordinate Reference System.

3.2.5.3 Geographic identifiers (ISO 19112)

The position of geographic features is often described through their spatial relation to other geographic features. This relation may be a containment, a local measurement, or a loose relation. An example of a containment is a city within a province; an example of a local measurement is the distance to the next major road intersection, and an example of a loose relation is a restaurant "between the museum and city hall". A typical application is the partitioning of an area using postal codes. These types of positions are addressed in the ISO 19112 (Spatial referencing by geographic identifiers).

All positions are related to a spatial reference system. The spatial reference system comprises a subdivision of a territory such as the hierarchy "province – city – address". The core element of the reference system is a gazetteer that adds a descriptive position to every geographic feature in the territory.

A gazetteer is a file that contains a master record for every geographic feature and the related descriptive position. If required, any descriptive position can be related to coordinates according to ISO 19111 (Spatial referencing by coordinates). The coordinates may be expressed as point coordinates or as a bounding box for curves or surfaces.

A location type according to ISO 19112 is a territorial unit of the spatial reference system. Examples of location types are an administrative area, town, locality, street, and property. A geographic feature in this context is called a location instance. A geographic feature is listed in the gazetteer and is related to one or more location types. For example, the city hall is a geographic feature that has one record in the gazetteer and this record contains a position that may be expressed in three ways: as an address, or as a containment in the city, or as a containment in the province.

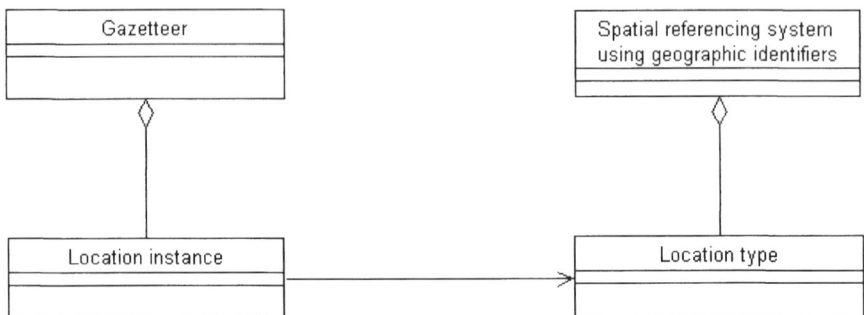

Fig. 3.13. Spatial reference system using geographic identifiers (ISO44 2001)

3.2.5.4 Positioning (ISO 19116)

Positioning services are computer techniques that deliver positions of an object relative to the earth or to some other position. These techniques include Global Navigation Satellite Systems (GNSS), Inertial Systems, and Total Stations. The ISO 19116 (Positioning Services) standardises a data model for the basic information independent of the system type, and a group of operations to handle those data. A system-specific section is dedicated to Global Navigation Satellite Systems only.

Finding a position is no longer the domain of a skilled navigator or of an experienced surveyor. But whomever the user, they can still suffer from the great variety of different and incompatible interfaces among the positioning systems. The ISO 19116 standardises this interface and isolates the client from the multiplicity of protocols.

Within the ISO 19100 family, the positioning services are among the processing services identified in ISO 19119.

Figure 3.14 gives an overview of the application cases of the positioning services. The diagram covers the standardised types including some future conceptual types.

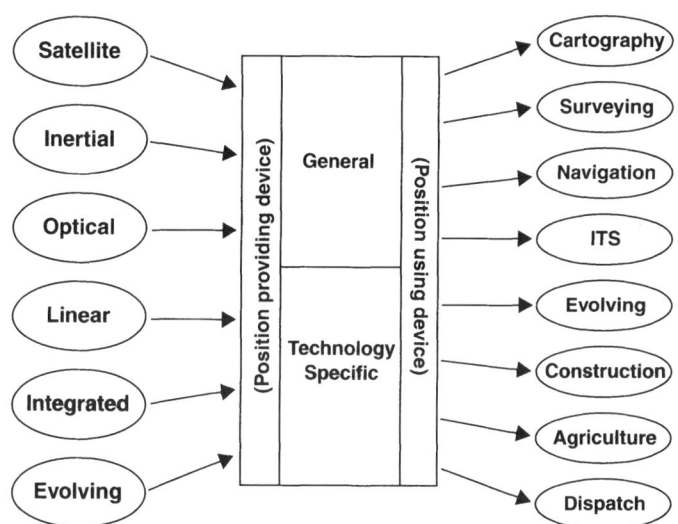

Fig. 3.14. Overview of application cases of the positioning services (ISO48 2002)

Figure 3.15 Shows the hierarchical structured data model for the basic information. The position information refers to the ISO 19111 (Spatial referencing by coordinates). The "Link to spatial referencing system" means a control point or a similar link between the instrument coordinate system and the earth in order to control accuracy aspects.

84   3 Non-geometry standards

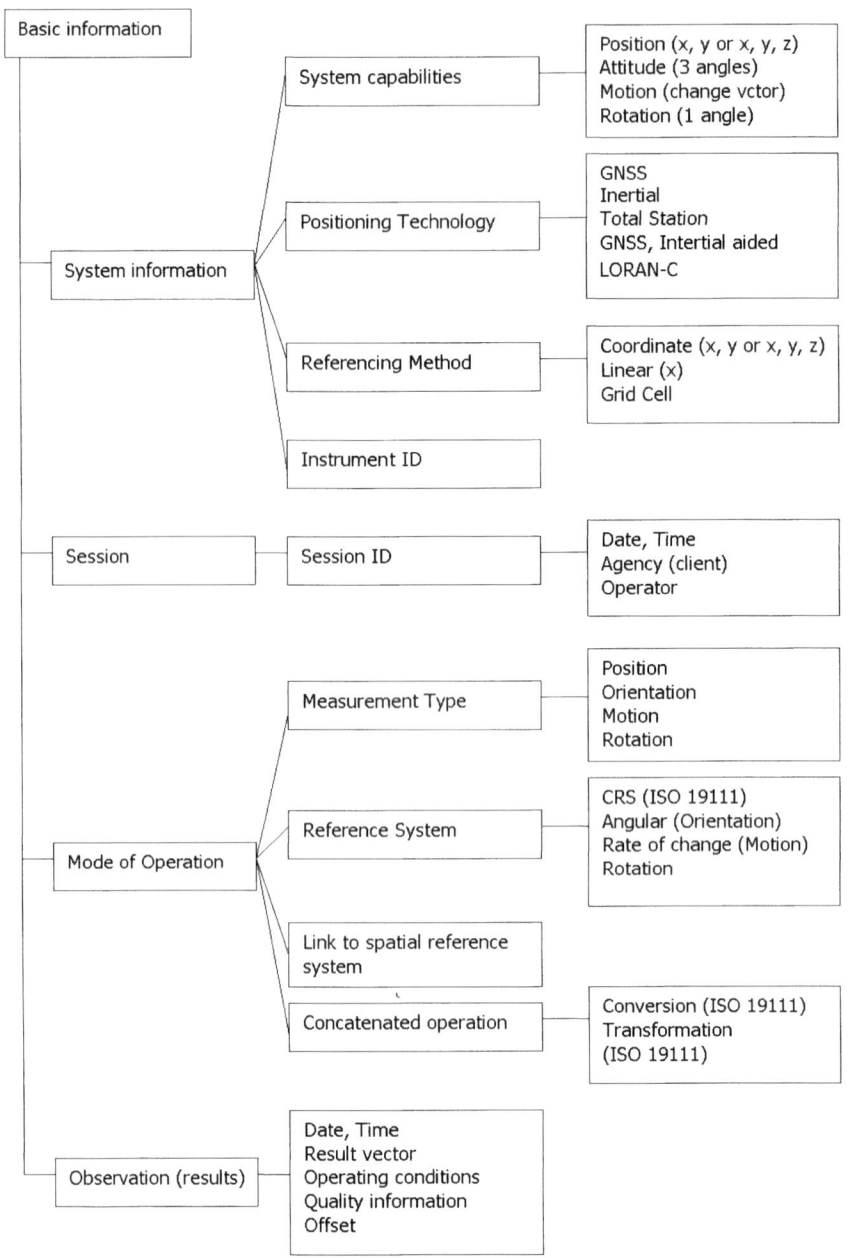

Fig. 3.15. Data model for the basic information of the positioning services

Morten Borrebæk is Norwegian. He studied surveying and photogrammetry at the Agricultural University of Norway in Ås near Oslo. For the longest time of his career Mr. Borrebæk was affiliated to the Norwegian Mapping Authority. He is presently the head of the section Geographic Information Technology.

Mr. Borrebæk started his career with research on digital geographic information at the Norwegian Institute of Land Inventory. Those works were focused on business applications in relation to the new technology.

Mr. Borrebæk was involved in GIS standardisation almost from the beginning. Already in 1985 he was coauthor of the Norwegian GIS standard SOSI. Later he belonged to the team that developed the European GIS standards (CEN/TC287) and subsequently he also has been a member of ISO/TC211 from the beginning. Even more, the Working Group 4 (Geospatial services) has been convened from its first day by Mr. Borrebæk.

3.2.6 Portrayal (ISO 19117)

Originally, the graphic presentation of geographic information was strictly the domain of cartography but has now also become an important section of geomatics. The ISO 19117 (Portrayal) defines a schema to create graphic output for datasets and metadata of the ISO 19100 family of standards. The scope of ISO 19117 does not include the standardisation of cartographic symbols.

According to ISO 19117, the cartographic symbolisation is kept separate from the feature types of the dataset. The definition of the cartographic representation for a feature is stored in a portrayal catalogue. Essentially the catalogue is a reference list that relates each feature code that is used to identify different feature types, to an individual cartographic portrayal.

A portrayal rule determines, which symbol shall be selected for a given feature. The rule may be simple such as a "black solid line" to portray a local road. Or a rule may be sophisticated such as a "double dashed red line" to portray a major road that carries 10,000 or more vehicles daily and that is maintained by the provincial government.

The ISO 19117 defines a two-step, rule-based approach to select a graphic representation for a portrayed feature type. In the first step, the available "rules" are tested according to a given feature type and its accompanying parameters. In the second step, the portrayal specification is retrieved in order for the rule to be found valid. A rule may be written as the SELECT-statement of SQL (Structured Query Language) and it contains one or more search criteria based on a parameter combination that may include map type, feature type, feature class, or scale range.

Examples of topographic feature types are streams, fields, and elevation points. A portrayal catalogue contains one cartographic representation for each feature type. For these examples it may contain a blue line for streams, a green fill-area symbol for fields, and a brown point for the elevation points.

The portrayal specification contains all the attributes and operations required to derive the graphic representation for the given feature type according to the applicable rule. The catalogue will always specify a default representation if the rule-search for a given feature type fails.

In order to adapt the graphics to any given symbolisation catalogue, external functions may be applied to the individual rules or to the whole dataset. An example is a car navigation system that will always display the map with the driving direction up.

Dense graphics may result in an uncontrolled overlay of the elements. A priority attribute allows defining the top element that hides the others in the case of multiple elements at the same position.

More than one feature catalogue may be defined for one dataset in order to allow for different types of maps.

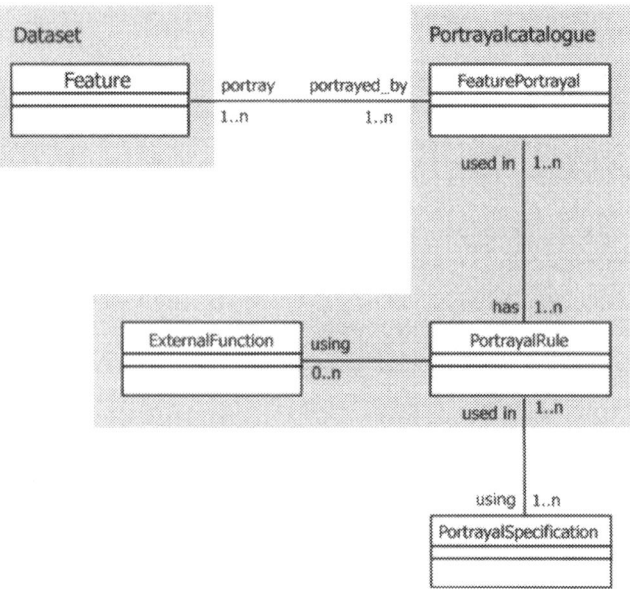

Fig. 3.16. Portrayal of a feature using a portrayal catalogue (ISO49 2002)

3.2.7 Encoding (ISO 19118)

The concept of the exchange of geographic information datasets is standardised in the ISO 19118 (Encoding). It defines a system-independent data structure for transport and storage and recommends the usage of the Extensible Markup Language (XML) for encoding. XML shall become a normative part of the ISO 19118 as soon as another standard of the ISO 19100 family, the ISO 19136 (Geography Markup Language), is accepted as mature.

The primary goal of the ISO 19100 family of standards is to enable full interoperability between heterogeneous geographic information systems. To achieve this goal, two fundamental issues need to be resolved. The first issue is to define the semantics of the content and the logical structures of geographic data. This is achieved by implementing the same application schema. The second issue is to define a system and platform independent data structure that can represent data corresponding to the application schema.

For example, equal semantics and logical structures can be guaranteed by using the same feature catalogue in the two systems involved. A system and platform independent data exchange format has to be created according to the ISO 19118 (Encoding).

Model for the data exchange

The generic model for the data exchange assumes two systems that run on different computer platforms and have different application schemas. The transfer of a dataset between such systems is modelled in six steps.

Step 1: The source system translates its internal data into a data structure that is according to the common application schema.
Step 2: An encoding service applies the encoding rules to the data, creating a file or transferring the data to a transfer service.
Step 3: The source system invokes a transfer service to send the encoded dataset to the destination system.
Step 4: The destination system receives the dataset.
Step 5: The destination system applies the inverse encoding rule to interpret the encoded data.
Step 6: The destination system must translate the application schema specific data into its internal database.

Steps 2 and 5 are standardised in ISO 19118 that describes the encoding and the decoding of the dataset. These two steps are handled by an encoding service that is a software component that implements the encoding rule and provides an interface to encoding and decoding functionality.

The encoding rule specifies the data types to be converted as well as the syntax structure and coding schemes used in the resulting data structure, preferably an XML-document.

Steps 3 and 4 use general Information Technology transfer services to send and receive data.

Steps 1 and 6, the internal translations within the source and destination systems, are outside the scope of this standard.

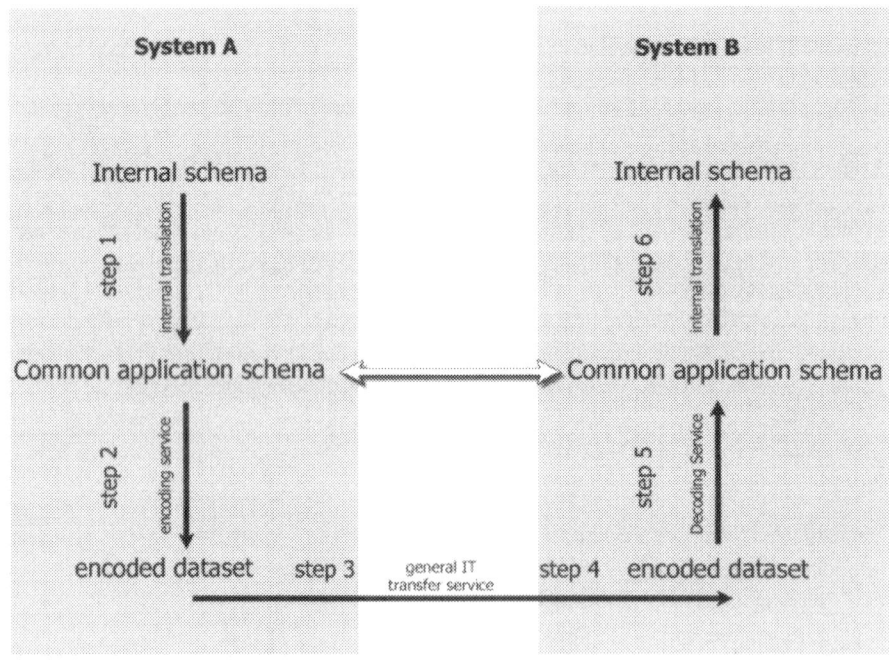

Fig. 3.17. Model for the data exchange, simplified (ISO50 2002)

Encoding using the Extensible Markup Language (XML)

As the Extensible Markup Language (XML) is recommended for the data exchange, the ISO 19118 (Encoding) defines a mapping for application schemas that are written in the Unified Modelling Language (UML) to the corresponding data structures of XML. There are two views to that mapping:

1. The *abstract view* is independent from any dataset. This is the *application schema*. It contains the classes and other components, such as attributes and associations, that fully describe all geographic feature types belonging to a certain application and their relations. A representation of the application schema is a feature catalogue. In XML the application schema is expressed as a Document Type Declaration (DTD) or as an XML schema document (XSD). Both contain the rules according to which an XML document has to be built.

2. The *implementation view* addresses a specific geographic dataset. This is the *application data*. It is, for example, a map compiled according to a feature catalogue. As every geographic feature is an *instance* of one class of the application schema, the dataset is addressed as the *instance model* in the ISO 19118. For the purpose of exchanging the dataset, the instance model is encoded in an XML-document. The structure and the tags of that XML-document are built according to the XML-DTD or to the XML-XSD.

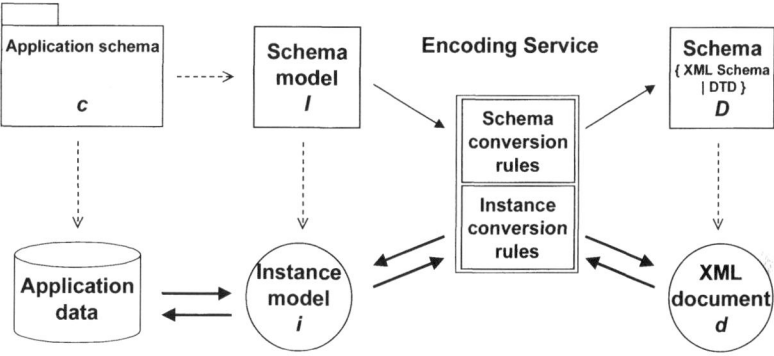

Fig. 3.18. XML-based conversion rules (ISO50 2002)

A more detailed introduction to XML is in Annex C.

3.2.8 Metadata (ISO 19115)

The ISO 19115 (Metadata) describes the metadata for documenting geographic information. It provides information about the identification, the extent, the quality, the spatial and temporal schema, the spatial reference, and the distribution of digital geographic data. This standard is applicable for describing geographic datasets, dataset series, and individual geographic features and feature properties. This standard can be used for cataloguing datasets, clearinghouse activities, and the full description of datasets.

The standard contains the complete listing of all metadata elements of the ISO 19100 family with a short explanation of each, however the detailed description of most elements can only be found in the individual standards. A complete listing of all metadata elements is included in this book (Annex B).

Metadata is a prerequisite to reuse geographic datasets. In a heterogeneous computing environment that has a variety of available data sources as well as a great number of different applications for the data, the metadata provide guidance to find the most appropriate dataset for a certain application. The metadata are capable of locating, evaluating, extracting, and employing the required datasets.

The current version of the ISO 19115 refers primarily to vector data. An amendment for imagery type data (raster) will follow (ISO 19115-2).

Content of ISO 19115

The metadata elements in ISO 19115 are grouped in two levels: the core metadata elements and the full list (also called "comprehensive metadata elements").

The core metadata elements are required to identify a dataset that is typically used for catalogue purposes. The core metadata answer the following questions:

- Does a dataset exist on a specific topic? (what?)
- Does a dataset exist for a specific place? (where?)
- Does a dataset exist for a specific date or period? (when?)
- Which is the point of contact that enables you to learn more about or order the dataset? (who?)

Table 3.5. Core metadata for geographic datasets

Core metadata elements	Remarks
Dataset title (M)	
Dataset reference date (M)	
Dataset responsible party (O)	
Geographic location of the dataset (C)	defined by four coordinates
Dataset language (M)	
Dataset character set (C)	
Dataset topic category (M)	
Spatial resolution of the dataset (M)	
Abstract describing the dataset (M)	
Distribution format (O)	
Additional extent (vertical, temporal) (O)	
Spatial representation type (O)	
Reference system (O)	
Lineage statement (O)	
On-line resource (O)	
Metadata file identifier (O)	
Metadata standard name (O)	
Metadata standard version (O)	
Metadata language (C)	
Metadata character set (C)	
Metadata point of contact (M)	
Metadata date stamp (M)	

(M) = mandatory, (O) = optional, (C) = conditional = mandatory under certain conditions

The metadata elements of the full list are packaged according to figure 3.19. For the most part, the packages relate to other ISO 19100 standard. For example, the "Data quality information" refers to the ISO 19113 (Quality principles), ISO 19114 (Quality evaluation procedures), and ISO 19138 (Data quality measures).

3.2 Basic standards

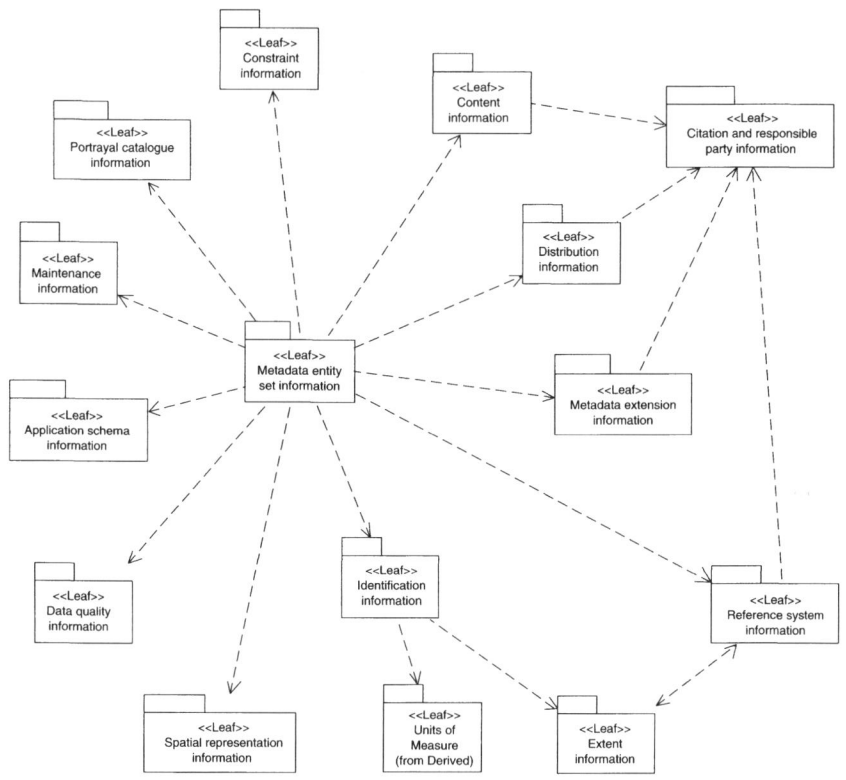

Fig. 3.19. UML package-diagram from ISO 19115 (Metadata) (ISO47 2003)

Concept of metadata

In the past, metadata was supplied as additional but separate pieces of information to a dataset. However, the concept of the ISO 19115 currently considers metadata as an integral part of the data that travels with data when copied, moved, renamed, or exported, and never gets lost. Metadata is implemented using standard technology like XML and large GIS manufacturers have started making the ISO 19115 metadata part of their system software (Danko 2002).

Hierarchy of metadata

Metadata standards existed before the work on ISO 19115 started as the development of the ISO 19115 intended to make it a superset of the existing metadata standards in the field of geographic information. A basic approach to metadata is the Dublin core (see below) but the most comprehensive existing source is the FGDC standard of the U.S. (Federal Geographic Data Committee). The design of ISO 19115 follows the ISO/IEC 11179 that addresses the specification and standardisation of data elements.

Work is currently continuing in an attempt to align the existing metadata standards to the ISO 19115.

The Dublin Core Metadata Element Set is a basic standard for the resource description in the IT-business and in particular, in library management. This standard was developed by the Dublin Core Metadata Initiative in cooperation with the NISO (National Information Standards Organization, USA), and approved by ANSI as ANSI/NISO Z39.85 – 2001. The Dublin Core defines only 15 metadata elements because simplicity lowers the cost of creating metadata and promotes interoperability. On the other hand, simplicity does not accommodate the semantic and functional richness supported by complex metadata schemes. The design of Dublin core mitigates this loss by encouraging the use of richer metadata schemes in combination with Dublin Core. The name Dublin Core refers to Dublin, Ohio, USA, where the first workshop of the initiative was held in 1995.

The technical background of David M. Danko is photogrammetry. In the very beginning he used to work in stereo compilation and aerotriangulation. Later he studied Geography and Cartography at the University of Maryland, USA. Today Mr. Danko is a Senior Consultant representing ESRI and the U.S. in national and international standards organizations.

Mr. Danko was employed with the U.S. federal government in the field of mapping for 39 years. As a physical scientist he was affiliated with the

U.S. Hydrographic Topographic Center, the U.S. National Imagery and Mapping Agency, and other agencies. One of his larger undertakings was a leading participation in the development of the Digital Chart of the World (DCW). Mr. Danko took responsibilities for securing funding, guiding concept development, prototype evolution, and standards development. The initiative included the United States, the United Kingdom, Canada, and Australia. This international effort resulted in the development of his first standard, the military Vector Product Format (VPF).

Metadata have become Mr. Danko's main concern. During his many years in this profession Mr. Danko has seen resources wasted, opportunities lost, miscommunication, the inability to adequately complete projects, and costly errors due to a lack of understanding of unfamiliar data and the lack of interoperability. Today Mr. Danko is one of the leading experts for geospatial metadata. Presently he is project team leader of the completed ISO 19115 (Metadata) and the new ISO 19139 (Metadata Implementation Specification). He also chairs the Open GIS Consortium Metadata Working Group.

Table 3.6. Comparison between the Dublin Core Metadata Element Set (Standard ANSI/NISO Z39.85) and ISO 19115 (Hudon 2002).

Dublin Core elements	ISO 19115 Core metadata compliance	ISO 19115 Comprehensive metadata compliance
Title	100 %	100 %
Creator	100 %	100 %
Subject	50 %	100 %
Description	50 %	100 %
Publisher	100 %	100 %
Contributor	0 %	0 % the meaning of "credit" is close
Date	50 % "creation date" included, "availability date" not included	50 % "creation date" included "availability date" not included
Type	0%	33%
Format	0%	50%
Identifier	0%	67%
Source	0%	0%
Language	50%	50%
Relation	0%	50%
Coverage	0%	100% the concept is called "spatial location"
Rights	0%	50%

3.2.9 Quality (ISO 19113, ISO 19114, ISO/TS 19138)

The value of geographic data is directly related to their quality. The views on quality differ among the various user communities. For example, cadastral applications require a positional accuracy within a few centimetres whereas a nautical chart requires a positional accuracy of only a few meters. Geographic datasets are being increasingly shared, interchanged, and used for purposes other than their producer's intended ones. The purpose of describing the quality of geographic data is to facilitate the selection of the geographic dataset best suited to application needs or requirements. The pair of standards ISO 19113 (Quality principles) and ISO 19114 (Quality evaluation procedures) define the principles for describing the quality.

Quality principles

The standards distinguish between data quality elements and data quality *overview* elements.

The "data quality elements" contain *quantitative* quality information. The ISO 19113 defines five groups to subdivide the elements:

Completeness	presence and absence of features, their attributes and relationships
	Negative example: missing road data in a remote part of the province.
Logical consistency	degree of adherence to logical rules of data structure, attribution and relationships
	Example: The application schema distinguishes between public and private buildings. The dataset distinguishes between low buildings and highrises.
Positional accuracy	accuracy of the position of features
	Example: The absolute point accuracy is 10 cm (diagonal).
Temporal accuracy	accuracy of the temporal attributes and temporal relationships of features
	Example: The date of the data compilation was August 1990.
Thematic accuracy	accuracy of quantitative attributes and the correctness of non-quantitative attributes, as well as the classification of features and their relationships
	Example: Areas have been classified according to remotely sensed imagery as green land although, in reality, they were swamps.

The data quality *overview* elements contain the non-quantitative quality information that is grouped in three:

Purpose	It describes the rationale for creating a dataset and contains information about its intended use.
Usage	It describes the application for which a dataset has been used.
Lineage	It describes the history of a dataset and, in as much as it is known, recounts the life cycle of a dataset from collection and acquisition through compilation and derivation to its current form.

Quality evaluation procedures

The ISO 19115 (Metadata) provides the dictionary for the data quality elements and the data quality overview elements. *Quantitative* and *non-quantitative* information can be reported as metadata according to ISO 19115.

According to ISO 19114, the *quantitative* information can also be reported in a "Quality evaluation report". The standard defines a detailed template for this report. There are two conditions under which a quality evaluation report shall be produced:

1. Data quality results are reported as metadata and can only be reported as pass-fail while more detailed information is needed.
2. When aggregated data quality results are generated the report explains how the aggregation was done and how to interpret the meaning of the aggregate result.

The figure 3.20 outlines the procedure to prepare the quality evaluation report.

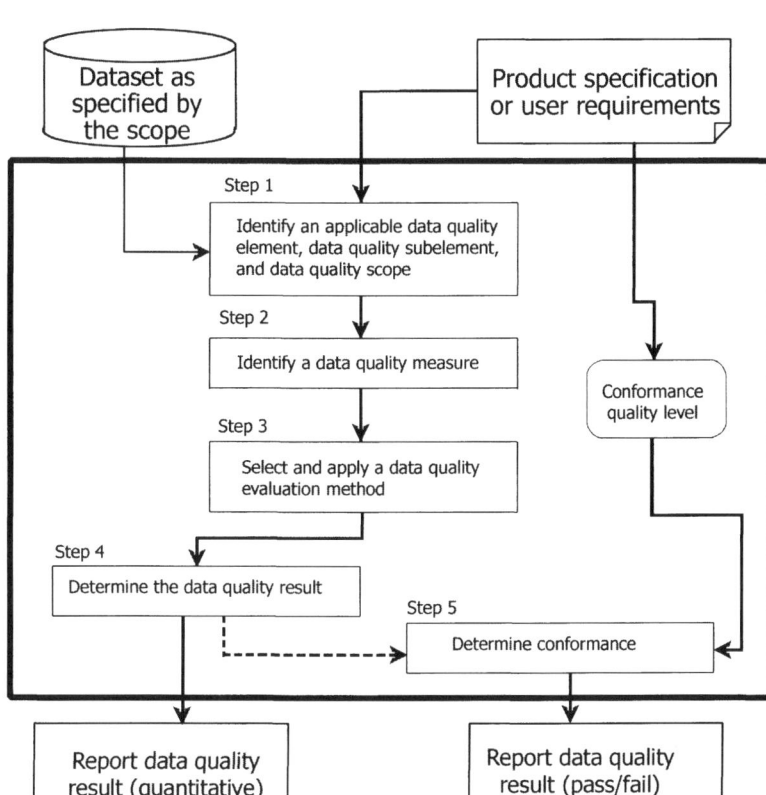

Fig. 3.20. Evaluating and reporting data quality results (ISO46 2001)

The ISO 19115 does not explicitly provide for the recording of quantitative quality information as metadata for feature *instances* or for attribute values. Quantitative quality information for single occurrences of items when differing from their parent types may be implemented by carrying the quality information as an attribute of the occurrences with the dataset.

Data quality measures

While writing this book, work is continuing on the ISO/TS 19138 (Data quality measures). This technical specification shall provide a definitive set of measures that can be used for different types of data in appropriate circumstances. The details of the measures shall be recorded in a registry according to the ISO 19135. The ISO/TS 19138 will enable the user to test a dataset against the measures in the registry. This specification is complementary to the standards ISO 19113 and ISO 19114 that define an *abstract* framework for the description of the quality of geographic data.

3.3 Imagery standards

The topic of imagery includes all geographic data that are stored in an image in the widest sense. The image may display a part of the earth's surface or it may contain other fairly evenly spaced data of the earth's surface or its neighbourhood. Examples for imagery-type data are airborne photographs, satellite images, hydrographic soundings, digital elevation models, and coverages according to ISO 19123.

3.3.1 Framework (ISO 19129, ISO/TR 19121, ISO/RS 19124)

When the work on ISO-standardisation of geomatics started, a great variety of raster data formats was in use already. Examples are JPEG and TIFF. Therefore, ISO standardisation for imagery started with a comprehensive review of industry- and other de-facto-standards. The results are a technical report (ISO/TR 19121), a review summary (ISO/RS 19124) and a framework standard (ISO 19129) that finally determines the further procedure. The concept of ISO 19129 contains the distribution of imagery elements to existing ISO 19100 standards, the definition of New Work Item Proposals (NWIPs) that will lead to new ISO 19100 standards, and eventually the standardisation of remaining topics within ISO 19129.

ISO 19129 (Imagery, gridded and coverage data framework)

The ISO 19129 is a framework standard for all coverage, imagery, and gridded data of geographic information. The term in the title with the broadest definition is "coverage". This term refers to geographic data that covers a region in some regular or irregular spatial order. The simplest order is a square grid, well known from digital images or digital elevation models. Thus, the concept of coverage data includes imagery and gridded data. A coverage is also considered as a kind of gridded data as well as a method for describing spatial phenomena that include concepts of vector and raster data. Coverages are explained in chapter 3, sections 3.4 and in chapter 4, section 5.

The ISO 19129 standardises all details that are necessary to handle gridded data in the environment of geomatics unless other standards of the ISO 19100 family have already done so. A close relation exists with the ISO 19130 (Sensor and data models for imagery and gridded data) that deals with the geometric relation of imagery data to the earth. Two "spin-off-standards" have already been defined, ISO 19101-2 (Imagery reference model) and ISO 19115-2 (Metadata – Part 2: Metadata for imagery and gridded data).

Enhancements of the existing ISO 19100 standards

The work on the ISO 19129 is still ongoing. The following listing explains additional important enhancements of the existing ISO 19100 standards for the needs of imagery.
- The ISO 19107 (Spatial schema) defines various surfaces such as a TIN. These surfaces will be referenced for imagery purposes.
- The geometric transformations of spaceborne imagery data often require coordinate systems to describe the orbit of a satellite. These systems are fixed in relation to the vernal equinox and are called Earth Centered Inertial (ECI) Coordinate Reference Systems. They may be added to the ISO 19111 (Spatial referencing by coordinates).
- The determination of the quality of imagery data requires visual and analytical inspection and evaluation methods. The quality of elevation models can be defined with measures distributed in a regular pattern across the model. This method is called "quality grid". The quality of imagery requires enhancements of the ISO 19113 and the ISO 19114.
- The portrayal of imagery data is often based on the application of Look-Up-Tables (LUT) or colour maps. The combination with vector-data often requires scaling and rotation of the image. These aspects lead to an enhancement of the ISO 19117 (Portrayal).
- A lot of ISO- and de-facto-standards exist for the encoding of imagery data. Hardly any standards exist for the storage of associated metadata. The ISO 19129 describes two ways for the neutral encoding of imagery metadata:
  1. General non-ISO picture coding and ISO-standardized XML for the metadata encoding.
  2. General non-ISO picture coding and ISO-standardized embedded metadata.
  An example for the first method is the TIFF-world format. An example of the second method is the TIFF-tags used in GeoTIFF. These requirements will lead to an enhancement of ISO 19118 (Encoding).

The following paragraphs describe tesselations that are not explicitly defined in the ISO 19123.

Tesselation is a method to order space. The ISO 19123 (Schema for coverage geometry and functions) contains a comprehensive definition of coverages. They comprise Thiessen polygons, quadrilateral grids, hexagonal grids, and TINs (Triangulated Irregular Network). For details see chapter 4, section 5. Additional tesselation methods are required for imagery applications.

The tiled tessellation partitions the space according to regional aspects without necessarily covering the complete area. It is used to partition a map series into single map sheets. It is also used to define outset of a map in order to bridge empty areas avoiding unnecessary data volumes. An example is a map of the U.S. showing Hawaii and Alaska on an extra tile.

A Riemann hyperspatial tessellation partitions a space according to the density of contents. Riemann has proven the applicability of the Euclidean geometry to the n-dimensional space. Thus the Riemann space is called a hyperspace. The Riemann hyperspatial tessellation is dynamic. Its basic partition is a multi-dimensional quadrilateral grid. While irregularly distributed data is stored in the dataset, the atoms of the hyperstructure are filled in with different intensity. Once the fill-grade of an atom reaches a threshold-value, it is then divided by two in each dimension. This means for example, that in the 3-dimensonal space a cube is split into 8 smaller cubes. The Riemann hyperspatial tessellation is widely used in domains with large amounts of irregularly spaced data like in hydrography.

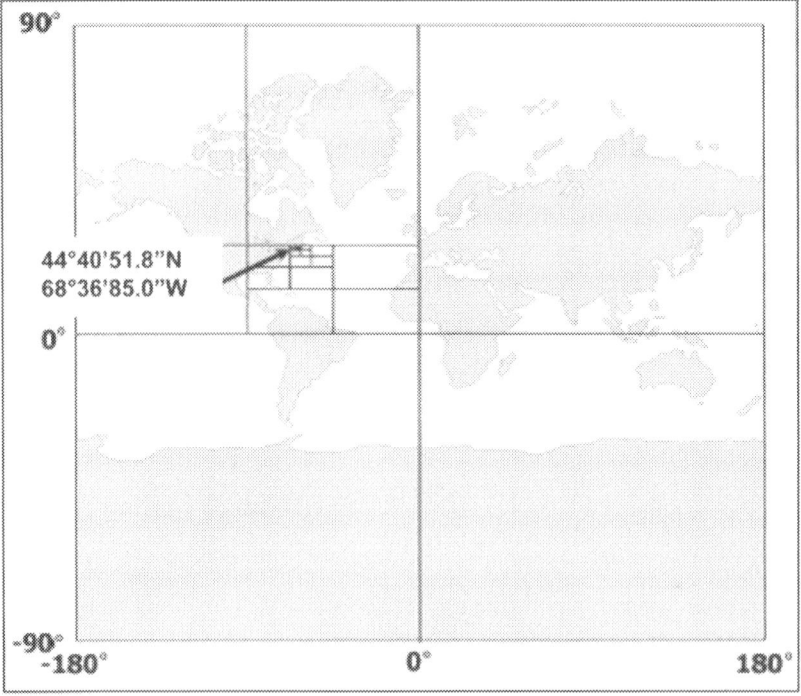

Fig. 3.21. 2-dimensional example of a Riemann hyperspace (Varma 2003)

An irregular grid is typical for swath-data. The scan-lines of swath data are not exactly parallel, because the sensor is in motion on a satellite- or on an aircraft-platform while the image is taken.

The ISO 19129 also covers systemic compression. It compresses data using content information. An example for a systemic compression is a topographic map where all information above water bodies is omitted.

Elements not referenceable to existing ISO 19100 standards

The radiometry is barely dealt with in the existing ISO 19100 standards. It includes the filtering of images, the histogram analysis, principle component transformation, and the adaption of neighbouring images.

Similarly the image analysis is not yet a part of the ISO 19100 family of standards. It includes edge detection, feature extraction, and image understanding.

Georeference

Georeference is the reference of a dataset to a location on the earth. Georeferencing information is the rationale of geographic information. From the procedural point of view some datasets are *georeferenced* once they are created while other datasets need to be georeferenced at some later time. These are called *ungeoreferenced*. Surveyed elevation points, small scale maps, or ortho photos are examples for georeferenced data. The raw data of remotely sensed imagery is an example for ungeoreferenced data.

Typically ungeoreference imagery data for geographic application has some additional information such as control points or orientation data that enable the process of georeferencing. These data are called *georeferenceable*. *Ungeoreferenceable* datasets, such as a portrait of a person, have no such additional information and are only of theoretical interest.

There exist two basic methods to establish the georeference of an imagery dataset: the *image reference* and the *sensor reference*.

The *image reference* uses image information and control points only. A geometrical model of the sensor that took the image is not required. This method has been widely applied for remote sensing imagery. The image reference can be established for instance with a 2-dimensional polynomial function that relates the image to the surface of the earth based on any number of common points called control points. The simplest way to georeference an image is the definition of the ground coordinates of the image corners. The image reference is considered an approximate method because terrain undulations between the control points can cause undetectable height parallaxes in the georeferenced image.

The *sensor reference* uses a mathematical model of the sensor such as the central perspective of a photogrammetric camera. This method requires a lot more initial parameters such as the focal length of the sensor and the position and attitude of the sensor while the image was taken. However, two or more images of the same region allow for a 3-dimensional data capture, for instance the creation of digital elevation model. A digital elevation model is required to create georectified orthophotos (see below) that are free from height parallaxes.

The sensor reference method is the domain of the classical photogrammetry. The advent of high resolution satellite sensors with metre-pixels on the earth's surface have stimulated a wider use of the sensor reference method in remote sensing. The sensor model is defined in the ISO 19130 (Sensor and data models for imagery and gridded data), chapter 3 section 3.2.

The most common geometry of georeferenced gridded data is the ortho projection. It denotes a vertical projection of the dataset to the reference plane like the xy-plane of a state Coordinate Reference System. Coverage-type datasets are based on points with coordinates like an elevation model or a thematic layer. If these datasets have 3 or more dimensions they are rectified by simply selecting two of the available dimensions, like the x and y. The result is a ground-plan of the dataset. This is the way maps are usually displayed.

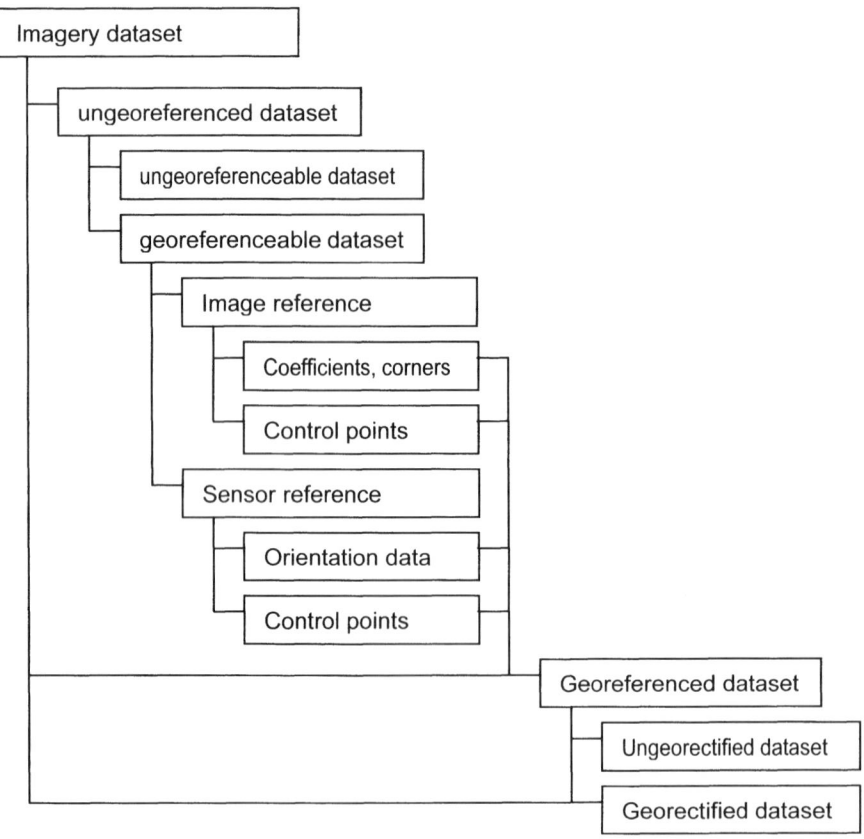

Fig. 3.22. Procedure of the georeference and the georectification for ungeoreferenced imagery datasets

Imagery-type datasets are always taken by a sensor, which mostly views the earth from a non-ortho perspective. For example, the photogrammetric camera has a central perspective towards the whole image and the line scanners have a central perspective across to the flight direction only. In these cases the sensors produce an inclined view of the earth - with the exception of the image centre – resulting in variable shifts of the image content according to the elevation of the ground. The process of changing the image content towards an ortho perspective is called orthorectification. This process requires a model of the elevation.

The ISO 19129 only deals with georeferenced gridded data. Ungeoreferenced data, if they can be referenced through the provision of control points, orbit parameters etc., are the domain of ISO 19130. All rules of ISO 19123 (Schema for coverage geometry and functions) assume georeferenced data like ISO 19129.

ISO/TR 19121 (Imagery and gridded data)

A survey of existing raster format standards was the starting point of the development of an ISO standard for gridded data. The result of the survey was the ISO/TR (Technical Report) 19121 that listed the following formats as being relevant for the upcoming standardisation. The ISO/TC211 has established formal liaison memberships or other appropriate links to most of the organizations responsible for the maintenance of the formats.

Table 3.7. Raster data formats identified in the ISO/TR 19121

Official standards (ISO, IEC, ITU, ISO/IEC JTC1), Government standards		Private standards	
Name	Maintenance	Name	Maintenance
BIIF	ISO/IEC JTC1/SC24	Coverages	OGC
CEOS CIP	CEOS	Fractal Transform	
CEOS Superstructure	CEOS	Coding	Microsoft
DIGEST	DGIWG (NATO)	GeoTIFF	JPL, Intergraph, Spot-Images
HDF	NASA (U.S.)		
HDF-EOS	NASA (U.S.)	GIF	CompuServe
IP-IIF	ISO/IEC JTC1	PhotoCampactDisk	Eastman Kodak
JBIG	ISO/IEC JTC1/SC29	TIFF	Aldus
JPEG	ISO/IEC JTC1/SC29		
MPEG-2	ISO/IEC JTC1/SC29		
NITF	NIMA (U.S.)		
PNG	ISO/IEC JTC1/SC 24		
S-57	IHO		
SDTS raster	ANSI (U.S.) and other countries		
SQL/MM	ISO/IEC JTC1/SC32		
T.4 and T.6	ITU		
TIFF/IT	ISO/TC130		

ISO/RS 19124 (Imagery and gridded data components)

The ISO/TC211 project 19124 (Imagery and gridded data components) is an intermediate step toward the completion of the ISO standards for gridded data and sensors. The result of the work is a "Review Summary" that is an informal ISO document that expresses the consensus of the project team members. The document reviews the existing standards for imagery and gridded data and analyses their impact on the ISO 19100 family of standards.

The project team decided to take this intermediate step because the development of de-facto standards in the field of imagery and gridded data was very advanced

102   3 Non-geometry standards

outside of ISO. In addition, the ISO/TC211 had agreed on the project 19123 (Schema for coverage geometry and functions) that is considered a superset of gridded data, and both created a situation that strongly required an alignment with existing solutions.

The existing standards listed in the ISO/TR 19121 were analysed using the following components: data model/schema, metadata, encoding, services, and spatial registration. These analyses resulted in a proposed amendment of the ISO 19115 (Metadata). The two other standards of the ISO 19100 family that require smaller amendments are the ISO 19104 (Terminology) and the ISO 19113 (Quality principles).

3.3.2 Sensors (ISO 19130)

Remotely sensed data is an important source for geographic information. Data are called remotely sensed if the sensor has no physical contact with the measured object. For the use of such data in combination with other geographic information, the remotely sensed data must be geometrically referenced to the earth. The ISO 19130 (Sensor and data models for imagery and gridded data) standardises the metadata for the *geometric* reference of the originally sensed data to locations on the earth. This reference allows for more precise data retrieval from the imagery. For most applications it is a prerequisite for an appropriate use of the imagery.

Sensor classification

From a conceptual view, the standard comprises sensor models for all remote sensing sensors. At present, the following sensors are described. They are grouped into the categories point type sensor, line type sensor, and area type sensor.

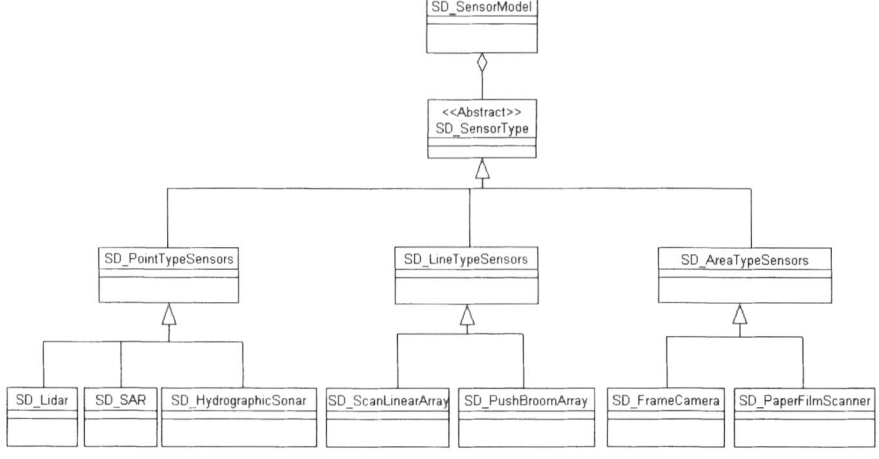

Fig. 3.23. Classification of sensors (ISO57 2003)

The *Lidar* sensor emits laser pulses and records the time until a reflection is received. It is mostly operated from an aircraft in order to determine the shape of the earth's surface. Airborne Lidar is often called *Laser Scanning*. Some systems include seabed surveys in shallow water.

Synthetic Aperture Radar (SAR) and Interferometric SAR (InSAR) type sensors are operated from satellite and aircraft platforms. The sensor emits radiation in the microwave band and records the time and the phase of the reflection. The system works under any light conditions (i.e., day or night). As microwave frequencies can penetrate clouds, the system's operation is not restricted to good weather conditions.

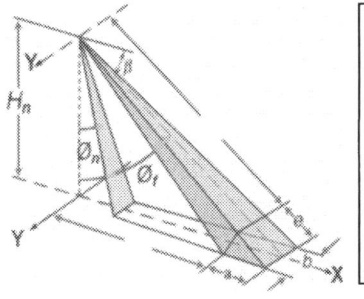

H_n = flying height
ß = depression angle
$Ø_n$ = near edge incidence angle
$Ø_f$ = far edge incidence angle
R_s = slant range
R_g = ground range
a = ground range resolution (x direction)
b = azimuth resolution (y direction)
e = slant range resolution

Fig. 3.24. Geometry of a side-looking Synthetic Aperture Radar (ISO57 2003)

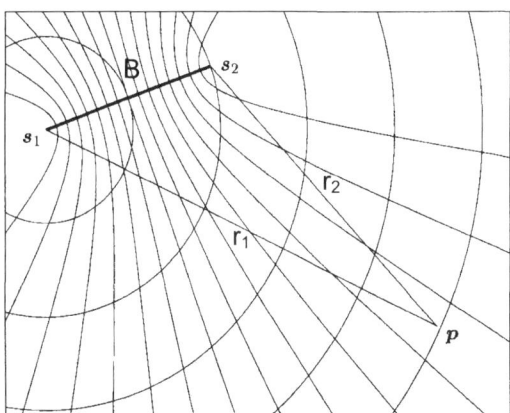

Fig. 3.25. Interferometric point positioning by range and phase difference
B is the base between two independent antennas. The circles correspond to "equi-range" lines to sensor s1 and the hyperbolas to "equi-difference of range" lines to both sensors. The flight direction is vertical to the drawing plane. (ISO57 2003)

The *hydrographic sonar* is used to measure the depth of the seabed and emits sound as well as measuring the time until a reflection is received.

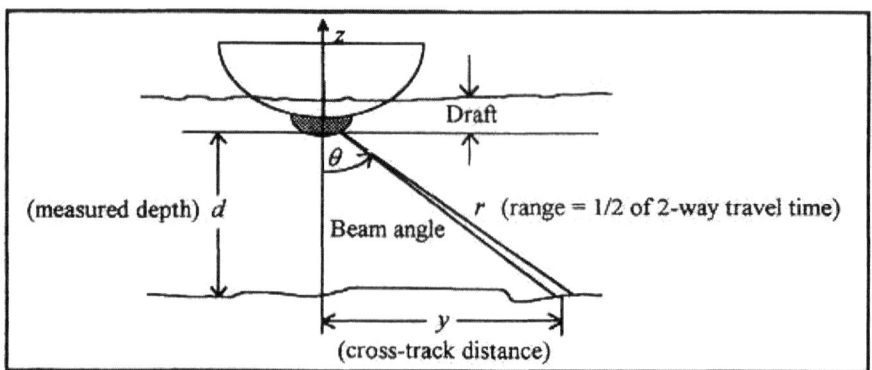

Fig. 3.26. Position and depth calculation for a hydrographic sonar (ISO57 2003)

The *scanning linear array* type sensor scans the terrain within the field of view under the flight track. This sensor scans along one scan line using a rotating mirror.

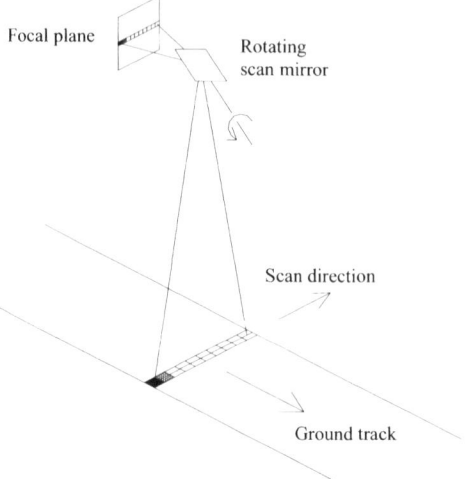

Fig. 3.27. Optical layout of a linear scanning sensor. The other types of scan linear arrays are conic scanning and whiskbroom scanning sensors

The *pushbroom sensor* takes the data along a scanline at one moment.

Most of the scanning linear arrays and pushbrooms are flown on satellite platforms. As the orbit is smooth compared to the ground resolution the resulting image may be georeferenced by robust transformations (sensor reference, rigorous model) or by functional relations such as polynomial transformations (image reference). If the scanning linear array type sensor is operated from an aeroplane, a line by line rectification is necessary to georeference the data.

3.3 Imagery standards   105

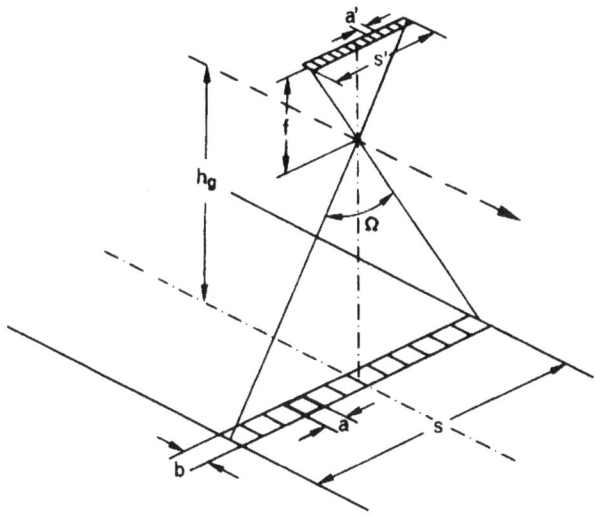

Fig. 3.28. Optical layout of a pushbroom sensor (Albertz and Kreiling 1989)

The *frame camera* produces a matrix of image pixels. Apart from some rare exceptions, the camera is flown on an aircraft.

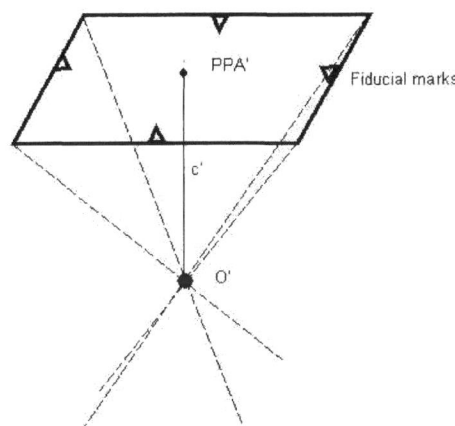

Fig. 3.29. Frame camera image and projection centre (ISO57 2003)

c'   Calibrated focal length, given in [mm]
PPA'   Principal point of autocollimation, given in [mm] in the image coordinate system
O'   Projection centre, given in [m] in the object coordinate system

Paper and film scanners convert paper or film originals to a grid values matrix.

Platform

The platform is the carrier of the sensor. The standard includes dynamic and stationary platforms. The dynamic platforms are spaceborne, airborne, landborne, or waterborne. The stationary platforms are landborne or waterborne. Each platform has its own characteristics in terms of position, attitude, and motion. For example, the track of a satellite is smooth and can be modelled as a Keplerian ellipse whereas the track of an aircraft is irregular and needs continuous monitoring of its position and attitude.

Georeference Models

For sensible applications ungeoreferenced imagery data need to be georeferenced. These imagery data are called georeferenceable. The ISO 19130 standard provides two models for the georeferencing of imagery to the earth.

The *rigorous sensor model* assumes a complete knowledge of at least the sensor geometry and the ground coordinate system to fully describe the involved Coordinate Reference Systems. It is based on the *sensor reference* to the object coordinate system. In many cases further knowledge (e.g. about the platform geometry and the orbit characteristics) is required too. The rigorous model allows for a point wise data capture with a defined accuracy. Stereo images enable the creation of 3-dimensional datasets.

Ground control points, the readings from Global Navigation Satellites Systems, and similar data sources carry the information about the ground Coordinate Reference System.

The *functional fit models* relate image to the earth by simply using a set of common points. The points define a functional relation, which is used to warp the original image towards the geometry of the earth surface. The model does not require any knowledge of the sensor. The functional fit model establishes an *image reference* between the remotely sensed data and the ground. The following functional fit methods are defined in the ISO 19130:

Ground control points

The *method* is called "Ground control points". The relation between the original image and the earth's surface is defined by a set of common points. Each point consists of two or three ground point coordinates (x, y, and optional z) and the related two sensor data coordinates (row and column). It is not specified which functional relation between image and earth shall be used and thus the type and the values of the coefficients remain undefined.

Grid interpolation

In this method, geolocation is derived by interpolation in an evenly spaced grid. Similar to the Ground control points method each grid point consists of three ground point coordinates and the related two sensor data coordinates. The ISO 19130 does not define the interpolation method to be used between the grid points.

Polynomials

The polynomials method presents the sensor data coordinate (row or track *i* and column or cross-track *j*) as a function of ground point coordinates (x, y, and optional z) in the form of a polynomial. For a polynomial, *i* may be written as

$$i = \sum_{k,l,m} a_{klm} x^k y^l z^m$$

and *j* as

$$j = \sum_{k,l,m} b_{klm} x^k y^l z^m$$

Ratios of polynomials

Rather than using a single polynomial, this geolocation method uses separate ratios of two polynomial functions of latitude, longitude, and height to compute image row and column. It is often called the Rational Functions model, and the polynomial coefficients are often called Rapid Positioning Capability (RPC) data. An image is divided into segments, and different polynomical ratios are used for the different sections. Each polynomial has 20 terms, although the coefficients of some polynomial terms are often zero. In the polynomial functions, the three spatial reference coordinates and two image coordinates are each offset and scaled so that their range over an image segment is -1.0 to +1.0.

For each image segment, the defined ratios of polynomials have the following form:

$$i_n = [p_1(x_n, y_n, z_n) / q_1(x_n, y_n, z_n)]$$

$$j_n = [p_2(x_n, y_n, z_n) / q_2(x_n, y_n, z_n)]$$

where
i_n = Normalized row index of pixel in image
j_n = Normalized column index of pixel in image
x_n, y_n, z_n = Normalized geographical coordinate values

The polynomials p and q have the following form:

$$p = \sum_{k=0}^{r_1} \sum_{l=0}^{r_2} \sum_{m=0}^{r_3} a_{klm} x_n^k y_n^l z_n^m$$

$$q = \sum_{k=0}^{s_1} \sum_{l=0}^{s_2} \sum_{m=0}^{s_3} b_{klm} x_n^k y_n^l z_n^m$$

where a_{klm} and b_{klm} are the polynomial coefficients.

The maximum power of each ground coordinate (r_1, r_2, r_3, s_1, s_2, and s_3) is 3. Furthermore, all polynomial coefficients where $k+l+m > 3$ are zero; the maximum order of any term is 3.

Universal real-time

The Universal real-time model is an extension of the ratios of polynomials or Rational Functions model, employing interpolation of high-order correction functions. Support data for this Universal real-time model is produced by fitting this model to any more rigorous sensor model, following completion of analytical triangulation or other adjustment. This method can be used for real-time image geolocation.

A georeferenceable image consists of the image itself, the specification of one of the above-mentioned methods, and all additional parameters necessary to compute the georeference.

3.3.3 Sensor Model Language (SensorML)

The Sensor Model Language (SensorML) is a framework for the description of sensor characteristics using the Extensible Markup Language (XML). The scope of the SensorML includes any sensors, remote sensing sensors, and in-situ sensors that receive their measurements through direct contact. This puts the scope beyond the domain of geomatics. There is however, a large overlap with the imagery standards of the ISO 19100 family. The SensorML might become an implementations of the ISO 19130.

The purpose of SensorML is to:

- Provide general sensor information in support of data discovery.
- Support the processing and analysis of the sensor measurements.
- Support the geolocation of the measured data.
- Provide performance characteristics (e.g. accuracy, threshold, etc.).
- Archive fundamental properties and assumptions regarding sensor.

SensorML provides an XML Schema for defining the geometric, dynamic, and observational characteristics of a sensor.

SensorML does not provide a detailed description of the hardware design of a sensor but rather it is a general schema for describing a functional model of the sensor.

SensorML supports both rigorous sensor models (sensor reference) and functional fit models (image reference). In the case of rigorous sensor models, the description of the sensor is typically separated from that of its platform.

Sensor Identification

The root for all SensorML documents is the Sensor and the sensor description is divided into main components of information, including:

- Sensor identification
- Document constraints like classification level or use restrictions
- Platform to which the sensor is attached
- Location of the sensor relative to a Coordinate Reference System
- Observable that can be measured
- Sensor's response characteristics like sensitivity
- Geometric and temporal characteristics of the samples
- Operator and Sensor Planning Service
- Sensor description such as serial number
- Documentation description such as author and modifications

Sensor models

Currently the SensorML describes the in-situ sensor case and three specific remote sensor models. These three models include (1.) the Scanner/Profiler Model, (2.) the Optical Camera Model, and (3.) the Rapid Positioning Coordinate.

1. Scanner/Profiler Model

   The *line-of-sight sensor* is the simplest profiler, consisting of essentially a single look direction. Examples of line-of-sight sensors include altimeters and range finders, ground-based atmospheric sounders, and non-sweeping Lidar.
   Conic scanners include those sensors that scan by sweeping around a rotation axis producing a cone pattern relative to the target body. Examples of conic scanners include SSM/I (1D) that sweeps around a single axis at a constant elevation angle.
   Line-scanners sample by sweeping around a rotation axis in such a way as they sample a linear array relative to the target. This is perhaps the most common sensor system used for lower resolution Earth observation. Examples are AVHRR and MODIS.
   The *imager* class of sensors can be considered a special case of line or conic-scanners in which the entire array of pixels is sampled at one instant in time. Examples are push-broom and CCD array sensors.

2. Optical Camera Model

This model includes *frame cameras*, *video cameras*, and *SLR cameras* (Single Lens Reflex cameras).

3. Rapid Positioning Coordinate

This model covers all functional fit (non-rigorous) models. It is based on the OGC Abstract Specification Topic 7: The Earth Imagery Case.

Platforms

The platform is the carrier for the mounted sensor and is of major concern for mobile sensors. Common platforms include ground stations, automobiles, aircraft, earth-orbiting satellites, ocean buoys, ships, and people. The SensorML distinguishes between the following platform types:

Stationary platform
Attached platform
Satellite platform
Aircraft platform
Land vehicle platform
Water vehicle platform

3.3.4 Coverages (ISO 19123)

The term coverage has different meanings in the geomatics community. It is often a term synonymous to a layer that is a thematic subdivision of a dataset. Historically, layers were transparent foils that represented one colour of a printed map. Within the ISO 19100 family of standards, coverages have an extended and more abstract meaning. Coverages are a concept considered to describe continuous and discrete spatial and temporal features, and thus integrating concepts of the worlds of vector and gridded data. The related standard is the ISO 19123 (Schema for coverage geometry and functions).

The term coverage, in the sense of the ISO 19100 family, has been adopted from the Open GIS Consortium.

Coverages may be discrete or continuous. An example of a discrete coverage is a map showing cities and their population where an interpolation between them would make no sense. An example of a continuous coverage is a temperature map that provides a temperature value for any position within the boundaries of a region.

Coverages may be 2-dimensional or 3-dimensional. Examples of 2-dimensional coverages could be a soil map and a digital elevation model where the soil type and the elevation respectively, are handled as attribute values. An example of a 3-

dimensional coverage is a 3D-grid with values of atmospheric or oceanographic parameters associated with both horizontal position and height (or depth).

Coverages may have a temporal dimension that is the third or fourth dimension in the cases of 2- or 3-dimensional coverages respectively. An example of a coverage with a temporal dimension is a dataset that contains the recorded daily temperatures of all weather stations of a region over a period of at least several days.

Space and the time are summarised as the spatio-temporal domain. The attributes belong to the attribute domain that is also called the range and a coverage may have multiple attributes at each position.

A coverage is a "world of its own" and is viewed as a geographic feature. A coverage includes the operations that are required to use its data which leads to the perspective that a coverage is a function that relates the spatio-temporal domain to the attribute domain. An example of such operations is a request for an attribute value at a given position.

The ISO 19123 standardises a number of specific coverages. It comprises the Thiessen polygon coverage, the quadrilateral grid coverage, the hexagonal grid coverage, the TIN, and the segmented curve coverage.

Chapter 4, section 5 "Schema for coverage geometry and functions" explains the details of the ISO 19123.

3.4 Catalogue standards

3.4.1 Procedure for registration (ISO 19135)

As the number of applications for digital geographic information grows rapidly, a centralised registration is a means of promoting compatibility and avoiding duplicate efforts. The ISO 19135 (Procedures for registration of geographical information items) specifies procedures for the registration of items of geographical information.

A registration is the assignment of an unambiguous name to an object and the registration authority keeps the registry. The parties submitting their application to the registration authority are called sponsors. An application may include a profile of the 19100 series of standards, a feature catalogue, or other specifications for geographic information. The items may, although it is not essential, be specified in an International Standard.

The standard allows more than one registration authority but they must be approved by the ISO. In order to allow the user of geographic information to keep an overview over the authorities involved, their number shall be kept as small as possible. The registration authority keeps the registry as an Internet database.

The content of a registry includes the following items:

- A unique numerical registration identifier
- A name for the registered object
- The name of the sponsoring authority
- The address of the sponsoring authority
- The date on which the object was entered into the register
- The registration status of the object that will have one of the following values "proposed", "accepted", "rejected", "replaced", or "obsolete"
- The date at which the registered object was superseded or made obsolete
- The registration identifier of the object that replaced it
- A description of the registered object, including the field of application
- A description of the relationship of the registered object to existing standards
- Information describing the registered object in compliance with the requirements of the technical standard that defines the object class

The registry allows a dynamic update of the registered objects.

3.4.2 Feature catalogues (ISO 19110)

The name of the ISO 19110 is "Methodology for feature cataloguing" with a feature being the fundamental unit of geographic information. The details of the model-

Hiroshi Imai is Japanese born in Kobe. He is a professor at the Department of Computer Science of the University of Tokyo. His research interests include algorithms, computational geometry, GIS, and optimization.

Dr. Imai holds a Bachelor, a Master, and a Diploma degree of the University of Tokyo in Mathematical and Computational Engineering. Later in his career he served as a visiting associate professor at the McGill University in Montreal, Québec, Canada, and as a visiting scientist at IBM, Watson Research Center in Yorktown, New York, USA.

Within the ISO/TC211 Dr. Imai is leading the Working Group 9 named Information management. He also was the project leader of the first standard of the ISO 19100 family to be completed as an International Standard, the ISO 19105:2000 (Conformance and testing).

Dr. Imai is a member of various national and international societies such as the Institute of Electrical and Electronics Engineers (IEEE), the Association of Computing Machinery (ASM) and the Information Technology Standards Commission of Japan (ITSJ).

ling of a feature and its relationships to other features are explained in the chapter 3, section 2.2 "Rules for application schema (ISO 19109)". A feature catalogue is compiled of features according to this definition.

A feature catalogue forms a repository for a set of definitions to classify significant real world phenomena to a particular universe of discourse. The catalogue provides a means for organising the data that represent these phenomena into categories in order to ensure the resulting information is as unambiguous, comprehensive, and useful as possible.

A feature catalogue has many purposes. The most important ones are listed below:

- A standard feature catalogue may be sufficient for many applications thus considerably reducing costs because the creation of a feature catalogue is a time-consuming effort.
- A feature catalogue should present an application-oriented particular abstraction of the real world represented in a given dataset and in a form readily understandable and accessible to users of the data.
- Often the feature types in different systems have equal or similar names like private houses and private buildings. A feature catalogue may serve to clarify where the classification differs.

The ISO 19110 provides a template for the organization of feature catalogue information. This template comprises sections for each feature type and the associated attributive information. A feature catalogue with only one feature is shown below in table 3.8 as an example.

Table 3.8. Example for a Feature Catalogue with one Feature Type (Bridge)

#	Feature catalogue element	Example data (Bridge)
	Feature Catalogue	
1	Name:	Small example feature catalogue
2	Scope:	Illustrate principle of feature cataloguing
3	Field of Application:	Help understanding the standard
4	Version Number:	1.0
5	Version Date:	2003-02-20
6	Definition Source:	none
7	Definition Type:	not applicable
8	Producer:	Authors of the book
9	Functional language:	not applicable
	Feature Type	
11	Name:	Bridge
12	Definition:	Raised structure built to support a road and serving to span an obstacle like a river
13	Code:	1001
14	Aliases:	
15	Feature Operation Names:	Opening periods
16	Feature Attribute Names:	Bridge availability, Bridge category
17	Feature Association Names:	Feature under the bridge
18	Subtype of:	Roads

Table 3.8. (cont.)

	Feature Operation	
21	Name:	Opening periods
22	Feature Attribute Names:	Bridge availability
23	Object Feature Type Names:	Road
24	Definition:	Road closed if bridge is closed
25	Formal Definition:	
	Feature Attribute	
31	Name:	Bridge availability
32	Definition:	Status for the ordinary use
33	Code:	2001
34	Value Data Type:	Boolean
35	Value Measurement Unit:	
36	Value Domain Type:	
37	Value Domain:	
	Feature Attribute	
31	Name:	Bridge category
32	Definition:	Type of transportation facility supported
33	Code:	2002
34	Value Data Type:	Character
35	Value Measurement Unit:	
36	Value Domain Type:	1 ("enumerated")
37	Value Domain:	
	Feature Attribute Value	
38	Label:	| Generic/unknown | Road | Pedestrian |
39	Code:	| 0 | 1 | 2 |
40	Definition:	| not applicable or | | pedestrians only |
		| impossible to determine | | |
	Feature Association	
41	Name:	Above-under condition
42	Inverse Relationship:	Other features below the bridge
43	Definition:	Guarantees that the bridge is always above the other features
44	Code:	3001
45	Feature Types Included:	Stream
46	Order Indicator:	0 ("not ordered")
47	Cardinality:	1 : ? = "One or more"
48	Constraints:	
49	Role Name:	

3.4.3 Data dictionary registers (ISO 19126)

The ISO 19126 (Profiles for feature data dictionary registers and feature catalogue registers) is a profile of the ISO 19110 and is a data dictionary describing the set of features and attributes that may be used to create a feature catalogue.

The ISO 19126 standardises the feature catalogue of the digital geographic information domain of the military in NATO countries, in particular the Feature and Attribute Coding Catalogue (FACC). The Digital Geographic Information Working Group (DGIWG) of NATO maintains it.

The ISO 19126 is linked to other standards of the ISO 19100 family, in particular to the ISO 19135 (Procedures for registration of geographical information items) and to the ISO 19106 (Profiles). The ISO 19126 may serve as a guide for other information communities to develop other compatible registers that may work in a system of authoritative referencing.

3.5 Implementation standards

3.5.1 Data product specification (ISO 19131)

The ISO 19131 (Data product specification) promotes the application of the ISO 19100 standards family for practical uses. A data product is a dataset containing data with a spatial relation to the earth. A data product specification is a formal document that defines the details of the data product for production, end-user, and other purposes. It is a precise technical description of the data product in terms of the requirements that it will or may fulfill. Therefore, purpose of the standard is to provide practical help in the creation of such specifications.

A good guideline for the content of a data product specification is listed in a normative annex to ISO standard "Data product specification". It helps to take all details into account without necessarily thoroughly knowing the structure of the whole ISO 19100 family.

3.5.2 Simple features (ISO 19125)

The ISO 19125 consists of two parts named ISO 19125-1 (Simple feature access – Part 1: Common architecture) and ISO 19125-2 (Simple feature access – Part 2: SQL option). The ISO 19125 is implementation-oriented.

Both parts deal with simple features. They are geometries restricted to two dimensions with linear interpolation between the vertices and may have spatial and non-spatial attributes. The simple features model consists of points, curves, and surfaces. The elements may be combined to geometry collections that include multi point, multi curve, and multi surface.

Part 1 of the standard defines the simple feature model. It is an abstract model in the sense that is independent from a specific computer platform. Part 2 defines a database access of simple features through an SQL-interface. The Open GIS Consortium has proposed the basic documents of the ISO 19125.

The more comprehensive standards such as the ISO 19107 (Spatial schema) and the ISO 19123 (Schema for coverage geometry and functions) have used some elements of ISO 19125-1.

The chapter 4, section 4 "Simple features" contains a detailed description of ISO 19125-1.

3.5.3 Web Map server interface (ISO 19128)

There are currently many mediums for communication being used but the Internet admittedly is the most important. If we use the Internet, we rarely think of the software running in the background. We take it as is. Our interface to the net is a browser. A great variety of maps or other visual representations of geodata have already become available on the Internet. A Web Map Service produces maps of georeferenced data. In this context, a map is a visual representation of geodata but it is not the data itself. The ISO 19128 (Web Map server interface) is a specification that standardises the way maps are requested by clients via the Internet and the way that servers describe their data holdings. The ISO 19128 is a web interface specification for mapping data.

Originally, the ISO standard "Web Map server interface" was a development of the Open GIS Consortium (OGC) (De La Beaujardière 2002).

Operations

The ISO standard "Web Map server interface" defines three operations, the first two of which are required of every Web Map server.

- Get Capabilities (required): Obtain service-level metadata that is a machine-readable (and human-readable) description of the Web Map server's information content and acceptable request parameters.
- GetMap (required): Obtain a map image for which the geospatial and dimensional parameters are well defined.
- GetFeatureInfo (optional): Ask for information about particular features shown on a map.

Processing

A standard web browser can ask a Web Map server to perform these operations by simply submitting requests in the form of Uniform Resource Locators. When invoking the operation GetMap, a Web Map server client can specify the information to be shown on the map. Some of the information that could be specified would include the number of layers, the "styles" of those layers, what portion of the earth is to be mapped (a "bounding box"), the projected or Coordinate Reference System to be used, the desired output format, the output size (width and height), and the background transparency and colour. When invoking GetFeatureInfo the client indicates what map is being queried and which location on the map is of interest.

According to the philosophy of the Internet, no main server is responsible for collecting the data requested by the client. The display data commonly reside only at the client's computer. A particular Web Map server provider in a distributed Web Map server network needs only to be the steward of its own data collection.

Because each Web Map server is independent, a Web Map server must be able to provide a machine-readable description of its capabilities. The "service metadata" enables clients to formulate valid requests and enables the construction of searchable catalogues that can direct clients to particular Web Map services.

A Web Map service may optionally allow the GetFeatureInfo operation. If it does, its maps are said to be "queryable", and a client can request information about features on the map by adding a position around which nearby features are searched for.

Cascading Web Map server

A "Cascading Web Map server" is a Web Map service that behaves like a client of other Web Map services and like a Web Map service to other clients. For example, a Cascading Web Map service can aggregate the contents of several distinct map servers into one service. Furthermore, a Cascading Web Map server can perform additional functions such as output format conservation of coordinate transformation on behalf of other servers.

3.5.4 Location based services (ISO 19132 – ISO 19134)

Since the advent of global navigation satellite systems people have been inventing new applications that combine the online-positioning data with other digital geographic information. The largest family of these products is called Location Based Services (LBS). The term is meant in the broadest possible sense because the sources of locations are not restricted to satellite navigation systems. The term includes any technology that delivers the position of a receiver online. For example, today it could be a cellular phone that derives its position from the geographic cell it is momentarily operated in. In one of the simplest applications the received positions are shown as a red moving dot on a digital map displayed on some hand-held computer. The positions might be used for further processing on the geographic data. The development of ideas and applications has just started.

The ISO is addressing the growing field of Location Based Services very carefully. This is in order to direct the developments towards compliance with the existing 19100 family of standards while at the same time avoiding any restrictions imposed by early standardisation as many new ideas are still emerging that could change the scope. ISO covers the field with three standards. These are ISO 19132 (Location based services possible standards), ISO 19133 (Location based services tracking and navigation), and ISO 19134 (Multimodal location based services for routing and navigation).

3.5.4.1 Trends and overview

The "Location based services possible standards" summarises the present trends in LBS:

- The Bluetooth consortium develops industry standards for the wireless communication. Their work will result in a change of attitude towards the Internet once it becomes fully available on mobile computers. In the near future, any geographic information that can be retrieved from the Internet will be available at any place where a mobile computer can be used.
- An efficient data selection for download has become one of the major concerns using the Internet. In the near future, local transmission stations will automatically send geographic data of regional interest for downloading on mobile computers.

For example, such a system called passer-by may be used for local traffic or tourist information.
- Computers will become lighter, cheaper, and easier to use. Presently, the so called personal digital assistant (PDA) holds all standard computer applications and fits into a trouser pocket.

The "Location based services tracking and navigation" and the "Multimodal location based services for routing and navigation" standardise route finding methods which is the most widely used application of LBS today. The subtopics might be summarised as follows:

- Find the best route in terms of travelling time or travelling cost between A and B.
- While travelling, find the best route if the conditions change along the route.
- Find the best route in terms of travelling time or travelling cost connecting multiple locations. This is classically named the "travelling salesman problem".
- Find the best route in terms of travelling time or travelling cost between A and B having multiple means of transportation available. The answer to this request is called a multimodal service.

The development of these standards is still ongoing.

3.5.4.2 Tracking and navigation

Tracking and navigation are the core applications of location based mobile services. They are covered by the ISO 19133 (Location based services tracking and navigation). Tracking is the process of following and reporting the position of a vehicle in a network. In some cases it may be the position of a hand-held computer only. Routing is the finding of optimal routes between positions in a network. Route traversal is the execution of a route, usually through the use of instructions at each node in the path, and a start and stop instruction, at the first and last position of the route. The combination of routing, route transversal and tracking is navigation. The optimal route is the one with minimal costs in terms of money, or time, or other parameters such as pleasure along scenic routes.

The basic assumption is that services made available on the Internet will be accessed by mobile devices in a manner similar to on-web clients, with the exception that the mobile client can update its own position during the process. For this purpose on-web proxy applications for the mobile client are required. These applications act as a device transformer for messages and data flowing between the web-service and the mobile client. The interface between the mobile client and the on-web programs is out of scope for this standard and is covered by standards written by and within ISO/TC204 (Intelligent transport systems).

Tracking

The tracked positions are defined by coordinates or other types of positional descriptions. A tracking service (class TK_TrackingService) delivers the positions, either one by one or as a sequential list.

The positions may be one of the types coordinate, place name, feature, linear reference, network, address or phone.

A trigger defines the moment or the location for delivering positional information. The triggers are generally of two types: triggered by an event, or triggered by the passage of time.

A transitional trigger delivers a new position dependent on the movement of the vehicle being tracked. Usually, events take place after the completion of a distance, or after a change of direction. The periodic trigger is used to control location sequences by setting temporal limits on how far apart in time tracking samples are taken.

The tracking location metadata include the mobile subscriber and the quality of positions. The mobile subscriber is the item being tracked such as a car with a navigation system. The quality of positions may be expressed in five different ways.

- Point estimate circle
- Point estimate ellipse
- Point estimate arc
- Point estimate sphere
- Point estimate ellipsoid

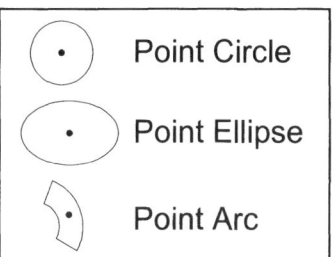

Fig. 3.30. Geometry of the point estimates circle, ellipse, and arc (ISO58 2003)

For position-type coordinates the tracking service uses linear reference systems that are in wide use in transportation. They allow for the specification of positions along curvilinear features by using measured distances from known positions, usually represented by physical markers along the right-of-way of the transportation feature. The ISO 19133 includes the description of positions close to but not on the path. The distance from a reference line to the path is called an offset. Figure 3.31 depicts a feature to the left of the berm of a road.

120 3 Non-geometry standards

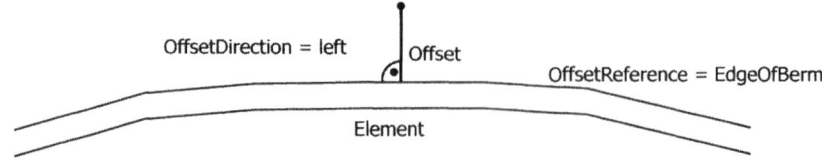

Fig. 3.31. Offset to a linear reference system (ISO58 2003)

"Element" is a part of the road's centreline that serves as the angular reference for the offset.

The linear reference systems are an extension to the Coordinate Reference Systems standardised in the ISO 19111.

In order to deal with address-type positions the ISO 19133 describes an address model. Currently there is not an applicable international address standard, and address formats vary from country to country and from culture to culture. According to the ISO 11180:1993 (Postal addressing) addresses are made up of a sequence of text elements with well-known semantic content. Table 3.9 lists the elements provided by ISO 19133 to specify an address.

Table 3.9. Elements of the address model of ISO 19133

Addressee	Phone number	Municipality quadrant
StreetIntersection	Named place	Region code (country)
Street	Street address	Number range
Postal code	Named place classification	List named places
Street location	Building	

Navigation

Navigational computation is based on an underlying network. The network comprises the following elements:

- The nodes represent important points in the network, such as intersections.
- The links represent uninterrupted paths between nodes with an orientation that indicates which direction the link is to be traversed.
- The turns associate a node to an entry link and exit link to a node.
- The stops consist of either nodes or positions on links within the network. The start and the end of the route are a type of stop.

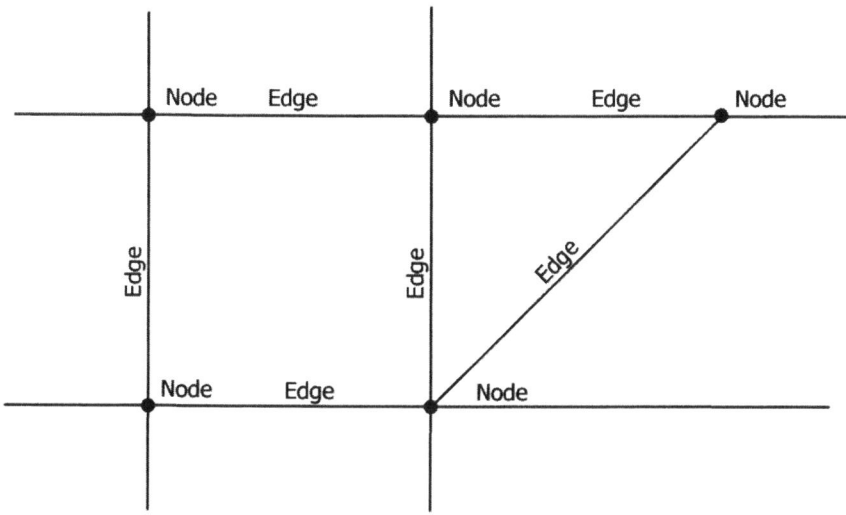

Fig. 3.32. Geometric topology of a road network

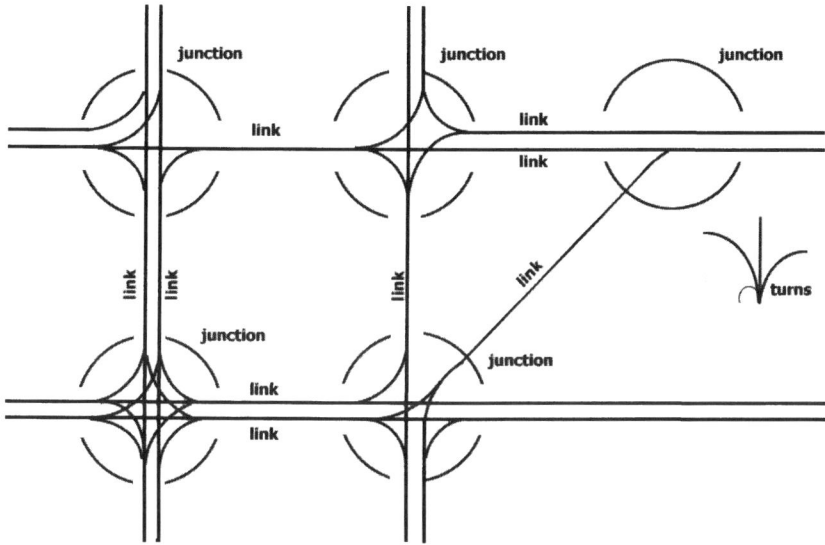

Fig. 3.33. Topology of a road network with links, junctions, and turns

A network has two different topologies. Its geometric topology refers only to nodes and edges as shown in figure 3.32. The second topology is the graph of links, junctions, and turns that is shown in figure 3.33. Although the links, junctions, and turns have the same underlying geometry they have their own connectivity based on usable "vehicle" routes. If a link comes to a cross-roads and U-turns are allowed then there are up to 4 turns which exit that link and enter into one of the links leaving that node including the one that reverses the incoming link.

The description of a network includes constraints such as vehicle constraints, temporal constraints, and lane constraints.

A navigation service delivers the optimal route between two positions within a network and guides the vehicle along that route.

A navigation service requests a route from the navigation system and receives a proposed route based on the given parameters. Table 3.10 shows the most important attributes for a route request:

Table 3.10. Attributes of a route request

Route request type	basic, predictive etc. (see below)
Vehicle	type such as car, or truck
Way point list	start point, end point, other stopping points
Avoid list	links in the network to be avoided
Departure time	planned period of the beginning of navigation
Arrival time	planned period of the end of navigation
Cost function	type, default is minimum distance
Preferences	specific user demands such as most scenic route
Advisories	text to be displayed during navigation
Refresh interval	maximum time before a recalculation of the route

According to their complexity the navigation services are categorized into five types:

- A *Basic navigation service* must support at least cost functions based on distance and expected average time.

- A *Predictive navigation service* is a basic service that must be able to take the chosen time of day and date into account for predicting travel time.

- A *Real time navigation service* is a predictive service that must be able to monitor traffic and road conditions and to reroute based on current information.

- A *Multiple stop service* is a basic, predictive or real time service that must be able to handle multiple stops (uncosted) along the route.

- A *Complex navigation service* is a multiple stop, real time services that must be able to include cost based on activities associated with traversal of the route, such as costing stops based on price of activities at those stops (see the description of cost functions below).

The cost function calculates the optimal route based on minimum costs. The ISO 19133 recommends the algorithms of Dijkstra (Dijkstra 1959) and Bellman-Ford (Bellman 1958; Ford 1956). Table 3.11 summarises the typical variables to control a cost function for car navigation.

Table 3.11. Cost function variables for route calculation

- Distance
- Time
- Stopping time (traffic lights)
- Speed
- Speed limits
- Slope (affects mostly freight vehicles)
- Link capacity (capacity of an intersection)
- Link volume (current amount of traffic at an intersection)
- Stopping time (expected stopping time at turns)
- Conditions (weather)
- Tolls

3.6 Qualifications and certification of personnel (ISO/TR 19122)

Beyond the scope of ISO/TC211, the ISO/Technical Report 19122 (Qualifications and certification of personnel) reviews the education and training for geomatics. The report primarily presents the result of a survey on the educational systems in 18 countries that responded to a questionnaire sent out to all members of ISO/TC211. A long-term goal of the effort is a plan for the accreditation of candidate institutions and programs and for the certification of individuals in the workforce. Within the geomatics community there is an ongoing discussion regarding the extent to which educational topics should be subject to a worldwide standardisation (Longley et al. 2001).

Qualification is the knowledge, skills, training and experience required to properly perform geomatics tasks that are normally achieved through formal education.

Certification is the procedure leading to a written testimony regarding the qualification of an individual's professional competence and can be provided by a range of public, private and professional institutions.

The intention of this informative ISO report (type 3 report) can be summarised in four statements:

1. Define the boundaries between geomatics and other related disciplines and professions.
2. Specify technologies and tasks pertaining to geomatics.
3. Establish skill sets and competency levels for technologists, professional staff and management in the field.
4. Research the relationship between this initiative and other similar certification processes performed by existing professional associations.

The survey questionnaire asked for the following information:

1. Does your country have a set of guidelines for the qualification and certification of personnel in the field of Geographic information/Geomatics?
2. If NO to Question #1, are you planning to initiate this activity in the near future?
3. Do you have national legislation for certification of personnel?
4. Do you have legislation for certification at the regional level?
5. Do you have industry standards?
6. Is there a group that has defined a model curriculum?
7. Do you have a mechanism for program accreditation?
8. How many higher education institutions teach Geographic information / Geomatics?
9. What Geographic information/Geomatics professional associations exist in your country?

18 countries responded to the questionnaire. In addition, 12 of them delivered a detailed case study on their national education in geomatics. The countries are Australia, Austria, Canada, Finland, Germany, Japan, Korea, New Zealand, Portugal, Saudi Arabia, South Africa, United Kingdom, and the United States. The FIG (Fédération Internationale des Géomètres) also provided a detailed report.

In a very general sense, the results of the case studies may be summarised as follows:

Japan and Korea have established national government bodies that have responsibility for the certification of personnel. Germany has a strong educational structure that has certain similarities to these Far East countries. In the UK, a number of professional organizations are linked to AGI (Association of Geographic Information) that is a consortium of private and public interests. Canada and the United States have academic consortia (e.g. UCGIS (University Consortium for Geographic Information Science)) and also active professional bodies. In South Africa a strong national association has formed to bring together regional GIS organizations.

4 Geometry standards

Traditionally, graphic data falls into the two categories of vector and raster. This approach is reflected in the way the ISO 19100 standards are partitioned. The ISO 19100 standards use the more general term "gridded data" instead of raster.

The ISO 19107 (Spatial schema) depicts a fairly complete world of vector data. The ISO 19125-1 (Simple features access – Part 1: Common architecture) is a subset of ISO 19107 using different terminology because of historic reasons.

The ISO 19123 (Schema for coverage geometry and functions) describes coverage data in a general sense. Coverages are often considered to be a way of integrating the worlds of vector and raster data thus overcoming the dichotomy originally imposed by hardware restrictions. However, coverages are restricted to three spatial and one temporal dimension. Later they shall handle an n-dimensional space. The ISO 19130 (Sensor and data models for imagery and gridded data) describes the models needed to relate remotely sensed imagery data to the earth. The ISO 19129 (Imagery, gridded and coverage data framework) is a framework standard that describes the concepts of gridded data, lists all required elements, and fills some gaps left over by the other standards.

4.1 Relations between the geometry standards

The following diagram shows the relations between the six geometry and imagery standards.

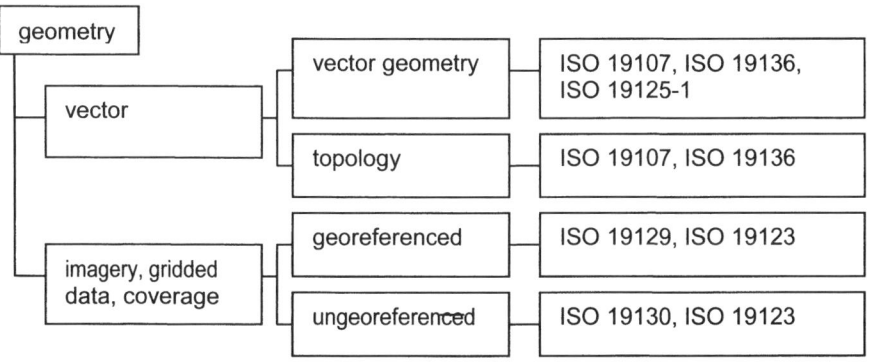

Fig. 4.1. Relation of the vector data and imagery standards of ISO 19100

Although a less abstract way of describing the structures may be desirable in some cases, the terminology of the conceptual schema language UML will be used in order to be consistent with the UML-models of the ISO 19100 standards. In most cases, the models consist of a hierarchy of classes. A high level class describes a graphical element in a very general sense. For example, a GM_Object can be any geometry object of the dataset like a point, a curve, a surface, or a solid. A low level class describes a specific geometry like an arc or a spline. The whole model contains an abstract view of the real world and it's only purpose is to address the classes that could possibly be present in a dataset. If a real dataset is created, it uses the classes supplied by the standard. The elements, which form the real dataset, are the instances of the abstract classes. For example, the instances of the class GM_Surface may be the property parcels 1, 2, …, n in the dataset.

4.2 Positions

The ISO 19107, the ISO 19123, and possibly others in the future have an integrated concept of dealing with positions in the n-dimensional space. The ISO 19125-1 supports only two dimensions and uses a different method.

The concept based on the ISO 19107 will be described first. An n-dimensional set of coordinates is called a DirectPosition like (x, y, z). Obviously a point object (class GM_Point) has one set of coordinates and thus one DirectPosition. Curves, surfaces, and solids have more than one set of coordinates and consequently more than one DirectPosition.

The name DirectPosition is derived from the *direct* identification of the position. Alternatively a position can be defined by the position of another "point object" (class GM_Point). This is called an *indirect* identification of a position. An example would be a parcel boundary which is mostly a line string. The positions of the line string are the corner stones of the parcel which are point objects themselves. Thus, the coordinates of the line string are the coordinates of the point objects.

In order to better handle numerous DirectPositions in a dataset, the positions of a curve, a surface, or a solid are collected in an ordered manner that is called a point array (class GM_PointArray). In order to avoid redundancy and unnecessary use of storage volumes, the point array only stores pointers and no coordinates. The pointers may point to DirectPositions or point objects.

The pointers within a point array (class GM_PointArray) are alternatively called controlPoints.

The class storing the flag with direct or indirect reference is called GM_Position.

The ISO 19123 concept of a coverage relates a set of attributes to a position within a bounded space. From this perspective, the basic element of a coverage is a pair of data collections; one data collection denotes a position like (x, y) and the other data collection denotes the attributes at this position like temperature and soil type. The ISO 19123 refers to the ISO 19107 for the details of positions.

4.2 Positions

The pair of data collections at one point is called a point value pair (class CV_PointValuePair). According to ISO 19107, the position of a point value pair is always a point object. In other words, the DirectPosition of ISO 19107 contains the coordinates of a position like (x, y) and the point object (class GM_Point) uses this DirectPosition. The point value pair of ISO 19123 (class CV_PointValurPair) uses the point object of ISO 19107 for position information.

A common pattern of points is a quadrilateral grid. Examples of this would be a digital image and a simple type of digital elevation matrix. For the reasons of storage and computational efficiency, the quadrilateral pattern of points has some special terminology and structure. The grid addressed as a whole is called a "grid values matrix" (class CV_GridValuesMatrix). The intersections of the grid lines are called "grid points" (class CV_GridPoint). According to ISO 19107, the position of each grid point is always a "point object".

ISO 19107 states that a grid can also be defined as a quadrilateral pattern of simple point positions or point objects (class GM_PointObject), as in ISO 19123. The grid is called a "point grid" (class GM_PointGrid) and consists of a number of parallel placed point arrays (class GM_PointArray).

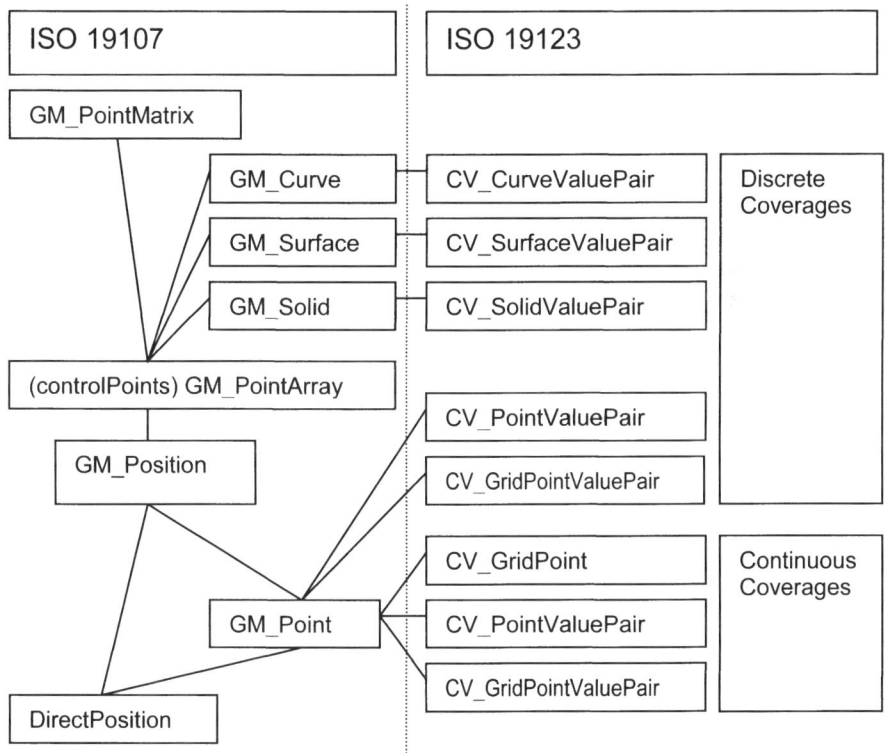

Fig. 4.2. Relations of position descriptions

A knot is a special type of point that is only used for the description of some types of splines. For example a Bézier spline is defined by four points where the curve passes through the first and last point and is only controlled in shape by the second and third point. The four points are the knots of the Bézier spline.

Each dimension requires its own sequence of knots. Thus a spline curve has one sequence and a spline surface has two sequences.

4.3 Spatial schema (ISO 19107)

The ISO 19107 covers a fairly complete world of vector data and has the following characteristics:
- 3-dimensional space, in theory n-dimensional
- primitives and complexes
- topological relations
- based on Set Theory

Set Theory is the mathematical science of the infinite. It is the study of the properties of sets, abstract objects that pervade the whole of modern mathematics. The language of Set Theory, in its simplicity, is sufficiently universal to formalize all mathematical concepts and thus Set Theory, along with a few other theories, constitutes the true Foundations of Mathematics (Jech 2002).

The ISO 19107 applies the axioms of Set Theory throughout the standard. An example is a geometric object that is defined as a set of geometric points.

The ISO 19107 has two important design criteria: the boundary-criterion and the complexes. The boundary-criterion means that high-level elements are composed of a collection of low-level elements. For example, a surface consists of its boundary curves, and the curves are bounded by their start and end points. The complexes, both geometrical and topological, consist of geometries, which do not overlap.

4.3.1 Overview over ISO 19107

The following diagram shows the logical tree of the important geometry classes of ISO 19107.

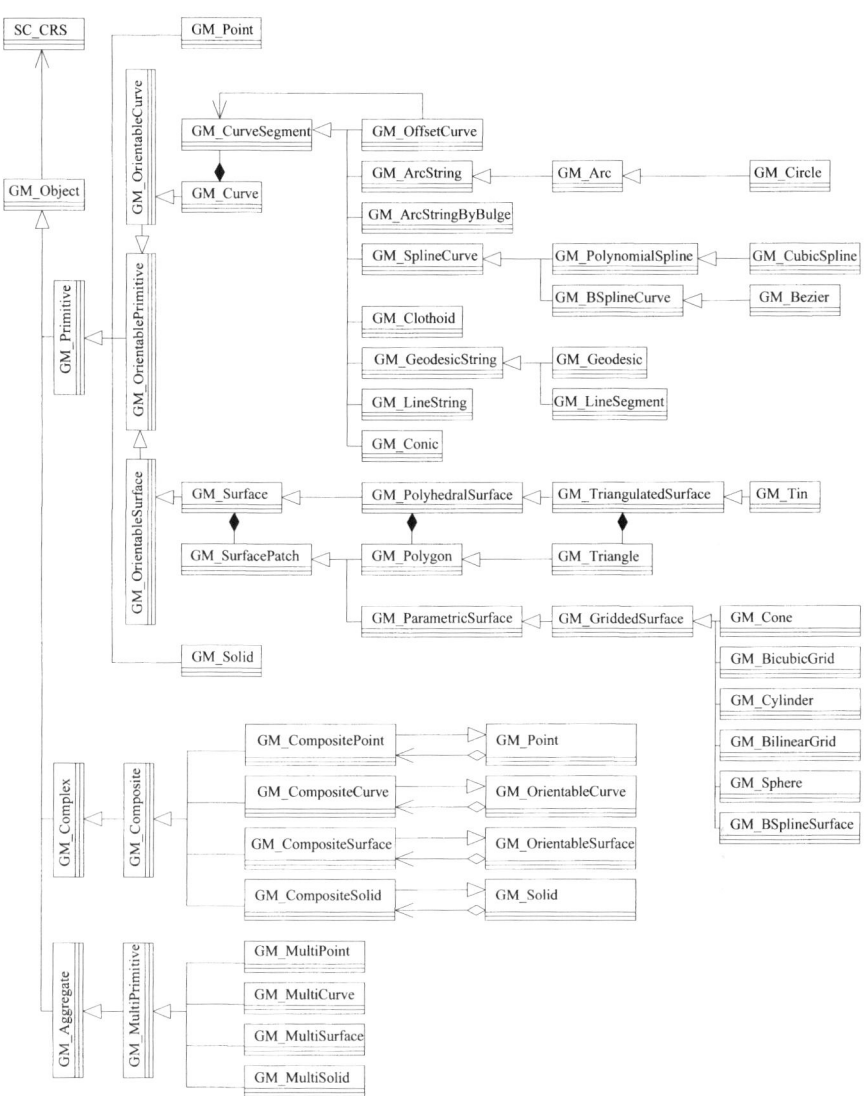

Fig. 4.3. UML class-diagram of GM_Object (ISO 19107)

4.3.2 General description of the geometry classes of ISO 19107

A geometry object (class GM_Object) is the most general concept for all objects that a geometrical dataset may consist of. On the top level, a geometry object may be a primitive (class GM_Primitive), a complex (class GM_Complex), or an aggregate (class GM_Aggregate). In every case the vector geometry of a geographic dataset can be completely described. In every case the fundamental geometries point, curve, surface, and solid are supported (classes GM_Point, GM_Curve, GM_Surface, and GM_Solid).

What is the difference between primitives, complexes, and aggregates? Primitives are the graphic elements that form the complete graphic of a geographic dataset. Primitives exist on their own and have no geometric relations to their neighbourhood apart from the common frame of a Coordinate Reference System. A typical map built on primitives and perhaps showing houses, roads, and rivers is primarily designed for visual information.

Complexes (class GM_Complex) allows the introduction of certain constraints among the graphic elements and are widely used in cadastral applications. In this prominent example, the parcels of a region have to cover the area without allowing any overlaps between neighbouring parcels and without leaving any gaps in between them. The components of a complex can be both a composite curve (class GM_CompositeCurve) and a composite surface (class GM_CompositeSurface). An example of a composite curve could be the boundary around a parcel. The term "composite" refers to the fact that a parcel usually has more than one neighbour and

John R. Herring, Ph.D., is American. His scientific background is mathematics. Today Dr. Herring does independent research and product design for spatial database products for the Oracle Corporation. His current work is both "heavily theoretical and heavily practical" in the application of the theory to spatial databases and applications.

Dr. Herring received a BA and a MA in Mathematics from the University of Alabama in Huntsville. He completed his Ph.D. in Mathematics on Differential Geometry and Topology at the Pennsylvania State University. For the last ten years he is Adjunct Professor at the University of Maine in Orono. Together with his colleagues in Orono he is editor-in-chief of the journal GeoInformatica.

Dr. Herring's involvement in standardisation began 1993 with SQL/MM. In ISO/TC211 he has edited the project 19107 (Spatial schema). Presently he chairs the project 19133 (Location based services tracking and navigation) and the Harmonized Model Maintenance Group that develops an integrated model for the whole family of ISO 19100 standards. He is engaged in ISO/TC211 because geographic information interoperability aids in Oracle Corporation database products for geospatial information. Dr. Herring also chairs Working Groups at the Open GIS Consortium.

thus the parcel boundary is composed of a number of single curves.

Aggregates (class GM_Aggregate) allow the grouping of geometric elements without any constraints. A typical example is a set of elevation points. Without aggregates being available the points could only be described as a number of individual points; the aggregate allows them to be addressed as a single dataset like a named list of points. Aggregations are named "multi primitives" such as multi points, multi curve, multi surface, and multi solid (classes GM_MultiPoint, GM_MultiCurve, GM_MultiSurface, and GM_MultiSolid).

Curves and Surfaces are orientable primitives (class GM_OrientablePrimitive) while points and solids are not. A curve is an orientable primitive because it has a defining point sequence that may have a forward or a backward order. This property is important if two or more curves form a closed polygon or a composite curve. A consecutive order of the points is required to correctly define the area as well as to draw a correct line pattern without interruptions along the curve where the individual curves meet.

A surface is an orientable primitive because either side may be the upside or downside (class GM_OrientableSurface).

A curve (class GM_Curve) consists of one or more curve segments (class GM_CurveSegment). A curve segment may have one of the following geometries: arc string, arc string by bulge, spline curve, clothoid, geodesic string, conic, and offset curve (see also figure 4.3 in chapter 4, section 3.1).

The surface (class GM_Surface) is either defined by a mosaic of surface patches with a great number of different shapes or by a simpler pattern of joint polygons. Surface patches might be parcels, areas of homogeneous land use, or local functional descriptions of the shape of a terrain. The perspective on surface patches focuses on every individual patch of the surface while the perspective on the pattern of joint polygons instead focuses on the surface as a whole. The most prominent example is a TIN (Triangulated Irregular Network) which results after processing a given set of elevation points. If one point of the dataset is changed then the whole TIN will change according to modifications of the triangles associated to the changed point.

The description of a complete surface with only one function is a rare case in geographic applications.

Under the restriction that every surface patch must be neatly linked to its neighbours in order to form a continuous surface without gaps, each patch (class GM_SurfacePatch) in a mosaic may have an individual functional description.

A simple case of a surface patch is a closed planar polygon (class GM_Polygon). A typical example is a parcel or a 2-dimensional feature in a land use dataset.

In ISO 19107 polygons are always planar. That means that all curves belonging to the polygon are part of the same plane. Geometrically this is always true for triangles only. Quadrangles and other polygons of higher order may have a 3D-shape. Those polygon are not valid polygons according to ISO 19107. If all surface patches are triangles being a special case of polygons, the result might be looking the same as a TIN. However, a pattern of triangles is a TIN only if the Delaunay criterion is valid as explained below (see GM_Tin in chapter 4, section 3.3.3).

The most sophisticated type of surface patch is a parametric curve surface (class GM_ParametricCurveSurface). A simple example is a semi-sphere. For practical reasons the defining points of a parametric curve surface often lie in a square pattern and thus form a quadrilateral grid. The resulting surface is called a gridded surface (class GM_GriddedSurface). The ISO 19107 standardises six types of gridded surface: cone, cylinder, sphere, bilinear grid, bicubic grid, and B-spline surface (classes GM_Cone, GM_Cylinder, GM_Sphere, GM_BilinearGrid, GM_BicubicGrid, and GM_BSplineSurface respectively). A bilinear grid uses line strings as horizontal and vertical curves while a bicubic grid uses cubic polynomial splines as horizontal and vertical curves. The names bilinear and bicubic grid will not be confused with the respective terms used for the interpolation methods.

A surface with only polygon surface patches is called a polyhedral surface (class GM_PolyhedralSurface). If the polygons are triangles it is called a triangulated surface (GM_TriangulatedSurface) with no restriction on how the triangulation is derived.

A TIN (class GM_Tin) is a triangulated surface that uses the Delaunay algorithm or a similar algorithm complemented with consideration for breaklines, stoplines, and maximum length of triangle sides. These networks satisfy the Delaunay criterion away from the modifications. For each triangle in the network, the circle passing through its vertexes does not contain the vertex of any other triangle in its interior.

The TIN is also covered by the ISO 19123 (Schema for coverage geometry and functions). The ISO 19107 standardises the description of an existing TIN. The ISO 19123 addresses the computation of a TIN and the interpolation of elevations.

4.3.3 Detailed description of the geometry classes of ISO 19107

The following chapter gives an exhaustive list of the geometric elements of ISO 19107.

4.3.3.1 Dimensions and map projection

The ISO 19107 standardises geometries in the 3-dimensional space as a base for the all ISO 19100 standards. The theory of ISO 19107 also allows its use in an n-dimensional space. However, the application of ISO 19107 leads to some inconsistancies.

Dimensions

The base geometries are points, curves, surfaces, and solids. Not all geometries are well defined in a 3-dimensional space.

Arcs and circles have their shape on one plane only. Any projection to a coordinate plane changes them to an elliptic arc or an ellipse other than the exceptional case that the plane of the geometry is parallel to a coordinate plane.

Cones keep their properties as being cones but change their shape.

The mathematics of splines and clothoids are defined in two dimensions only. In the case where they are used in three dimensions, the formulas are usually expressed in the parameterized form. This simply means that an additional dimension "t" is introduced. The origin of t is at the beginning of the curve and the value of t is the length of the curve from the beginning. Using t, splines and clothoids can be expressed on the three separate planes xt, yt, and zt (Foley et al. 1987).

An offset curve requires the additional information of which spatial direction the offset should be counted towards. This information is beyond the ISO 19107.

There are only a few geometries (such as lines and geodesics) that can be applied to an n-dimensional space without mathematics that are beyond the scope of ISO 19107.

Map projections

The geometries of ISO 19107 are built on the Euclidean space. The ISO 19111 standardises the coordinate transformations and conversions, including map projections. They usually change coordinates from the Euclidean space to some other space, like a sphere. With the exception of points, the map projection changes the shape of all geometries because the modified geometries do not have the characteristics of the original geometries. For example, a line becomes an arc after projection from a plane to a sphere.

4.3.3.2 Curves

GM_LineString

A line string (class GM_LineString) consists of a sequence of line segments.

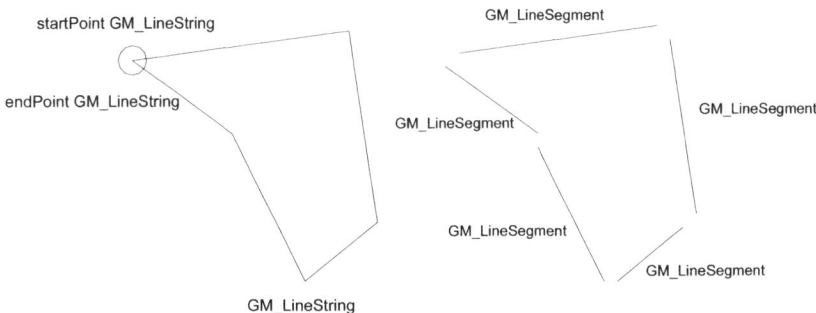

Fig. 4.4. GM_LineString

GM_LineSegment

A line segment (class GM_LineSegment) is a line between two given positions. This line will either be straight or some other shape depending on the map projection.

GM_GeodesicString

A geodesic string (class GM_GeodesicString) consists of a sequence of positions interpolated using a geodesic (class GM_Geodesic) defined from the geoid or ellipsoid of the Coordinate Reference System being used.

GM_Geodesic

A geodesic is the shortest line between two points on an arbitrary surface. In the trivial case of a plane, the geodesic is the straight line between two points. In navigation the geodesic on an ellipsoid is called the orthodrome.

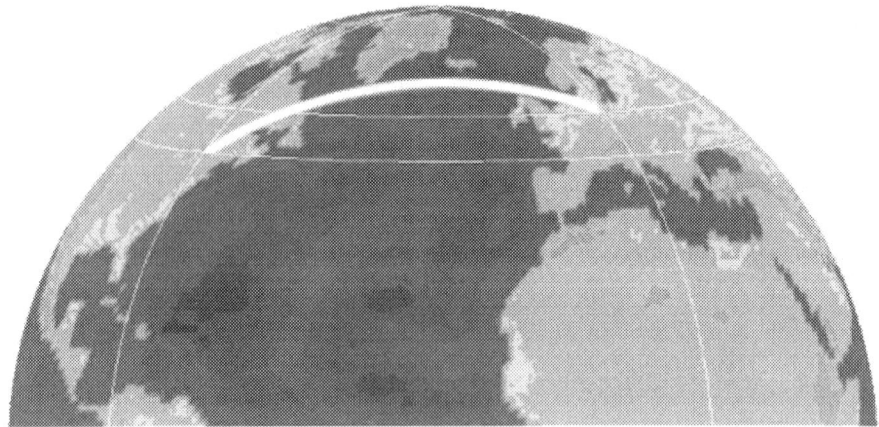

Fig. 4.5. GM_Geodesic: orthodrome between Ottawa and Neubrandenburg (Nitschke 2003)

GM_ArcString

An arc string (class GM_ArcString) consists of a sequence of arcs. Since it requires 3 points to determine a circular arc, the controlPoints are treated as a sequence of overlapping sets of 3 positions, the start of each arc, some point between the start and end, and the end of each arc. Since the end of each arc is the start of the next, this position is not repeated in the controlPoint sequence.

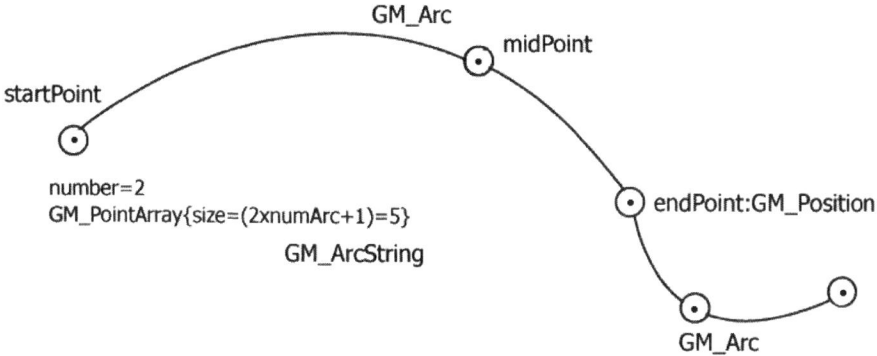

Fig. 4.6. GM_ArcString

GM_Arc

An arc (class GM_Arc) that consists of an arc of the circle is determined by three points and is therefore defined by the starting point, a point somewhere between the beginning and the end and the point of termination. If the 3 points are co-linear, then the arc shall be a 3-point line string, and will not be able to return values for center, radius, start angle and end angle.

GM_Circle

A circle (class GM_Circle) is defined by three points that close to form a full circle.

GM_ArcStringByBulge

An arc string by bulge (class GM_ArcStringByBulge) consists of a sequence of arcs where each arc is controlled by its startPoint, its endPoint, and a value determining its bulge. The bulge controls the offset of each arc's midpoint. The controlPoint sequence consists of the start and the end points of each arc. The bulge sequence consists of the bulge values, which is exactly one less than the length of the controlPoint array.

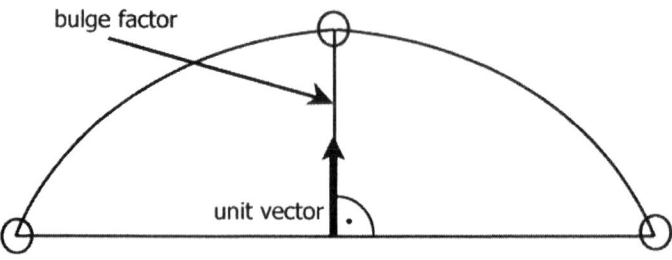

Fig. 4.7. GM_ArcStringByBulge

GM_Conic

A conic (class GM_Conic) may be an ellipse, a parabola, or a hyperbola depending on the parameter eccentricity. The given formula represents a conic in polar coordinates. The diagram shows four cases of different eccentricity.

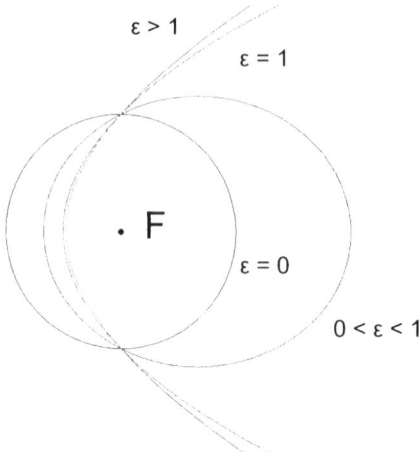

Fig. 4.8. GM_Conic (Nitschke 2003)

Polar coordinates: (ρ, φ) with $\rho = P / (1 + \varepsilon * \cos \varphi)$
$P = 1$
Eccentricity: $\varepsilon=0$ (circle), $0<\varepsilon<1$ (ellipse), $\varepsilon =1$ (parabola), $\varepsilon >1$ (hyperbola)

GM_Placement and GM_AffinePlacement

Placements are meant to take a standard geometric construction such as a circle, and place it as a symbol in a geographic dataset. In most cases a placement involves a shift of dimension from the 2D-space to 3D-space.

The placement (class GM_Placement) allows any type of geometric transformation to be applied during the output to the target system. With the transformation involved, a circle may be changed to any type of conic.

The affine placement (class GM_AffinePlacement) allows a linear transformation only (6 parameters).

GM_Clothoid

A clothoid (class GM_Clothoid) is a plane curve whose curvature is a fixed function of its length. The following equation is valid at any position along the clothoid:
length * radius = constant.
The clothoid is used to lay out roads and railway lines.

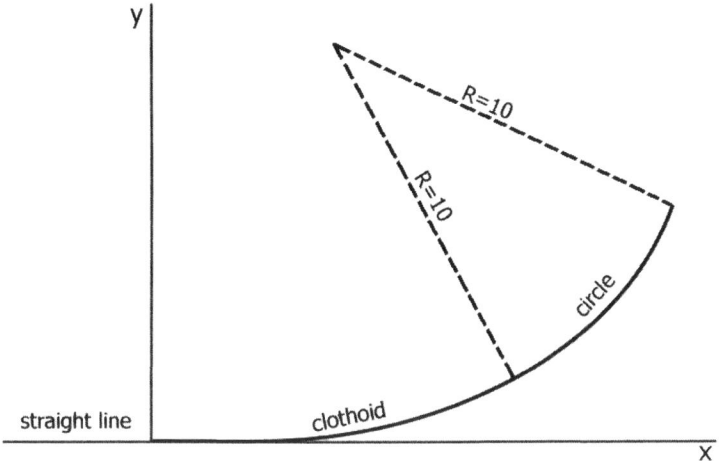

Fig. 4.9. GM_Clothoid (Larisch 2003)

GM_OffsetCurve

An offset curve (class GM_OffsetCurve) is a curve at a constant distance from the basis curve. They can be useful as a cheap and simple alternative to constructing curves that are offsets by definition.

GM_SplineCurve

Spline curves (class GM_SplineCurve) may be polynomial splines or B-splines.

GM_PolynomialSpline

An "n^{th} degree" polynomial spline is defined piecewise as an n-degree polynomial, with up to C^{n-1} continuity at the control points where the defining polynomial changes. A C^{n-1} continuity means that the original curves as well as the 1st, 2nd, ..., and $(n-1)^{th}$ derivative meet at the control points. Parameters shall include directions for as many as (degree - 2) derivatives of the polynomial at the start and end point of the piece.

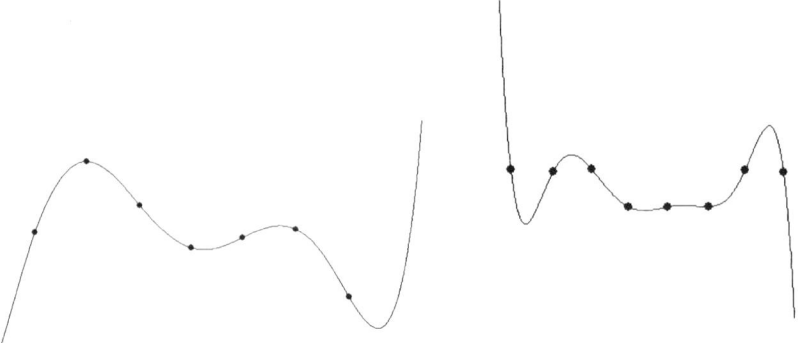

Fig. 4.10. GM_PolynomialSpline

The left example has the same points as the front profile in figures 4.18 – 4.20, 6th grade. The right example demonstrates strong oscillations that are typical for polynomials, 7th grade (Mak 2002).

GM_CubicSpline

A cubic spline (class GM_CubicSpline) consists of a sequence of segments each with its own defining function. A cubic spline uses the control points and a set of derivative parameters to define a piecewise 3rd degree polynomial interpolation.

The function describing the curve must have a continuous 1st and 2nd derivative at all points and pass through the controlPoints in the order given. Between each pair of the control points, a curve segment is defined by a cubic polynomial. At each control point, the polynomial changes in such a manner that the 1st and 2nd derivative vectors are the same from either side.

A special provision must be made for the first and last point of the spline because the tangent at these points remains undefined. The control parameters record must contain a vectorAtStart and a vectorAtEnd as these are the unit tangent vectors at controlPoint[1] and controlPoint[n] where n = controlPoint.count.

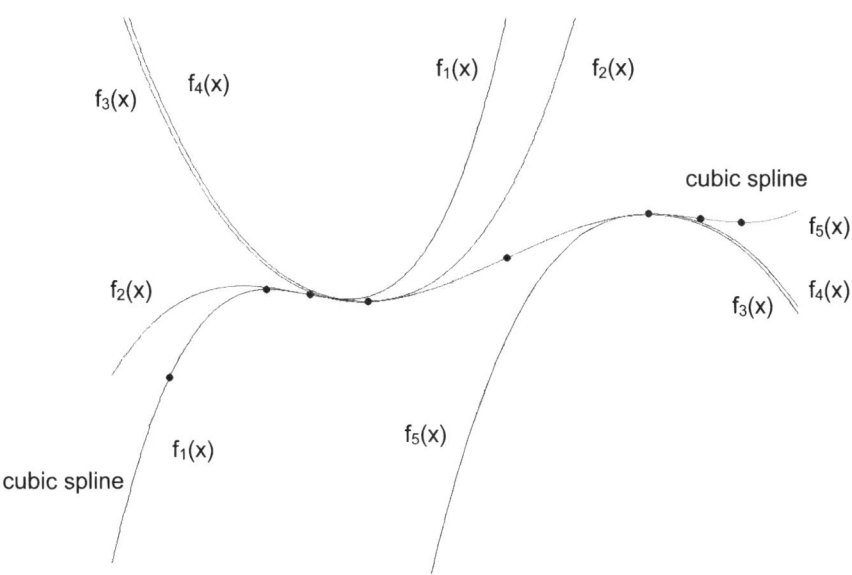

Fig. 4.11. GM_CubicSpline

Function 1: $f_1(x) = a_{01} + a_{11}*x + a_{21}*x^2 + a_{31}*x^3$
Function 2: $f_2(x) = a_{02} + a_{12}*x + a_{22}*x^2 + a_{32}*x^3$

Function 5: $f_5(x) = a_{05} + a_{15}*x + a_{25}*x^2 + a_{35}*x^3$
The first and the last segment are a part of the functions f_1 and f_5 respectively (Nitschke 2003).

GM_BSpline

A B-spline (class GM_BSpline) consists of a sequence of segments each with its own defining function, similar to the cubic spline. The shape of each segment is controlled by a weighted influence of the four control points next to the segment. The influence is described through four polynomials, one for each control point. The definition range of the polynomials is zero to one referring to the length of the segment. The polynomials are called basis or blending functions. A typical set of basis functions is the Bernstein polynomials in figure 4.12.

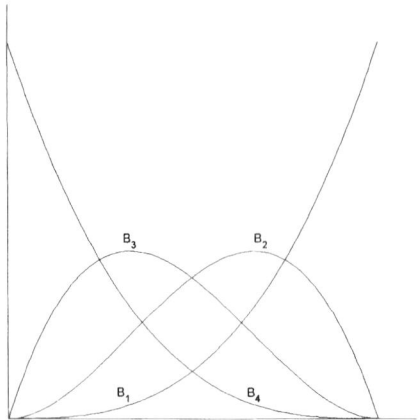

Fig. 4.12. Bernstein polynomials (Nitschke 2003)

The control points of a B-spline are called knots. If the knots are spaced in equal intervals the B-spline is uniform and if not, it is considered non-uniform.

If a B-spline is a polynomial then it is non-rational. However, if the B-spline is a ratio of polynomials it is then considered rational.

GM_Bezier

A Bézier-spline (class GM_Bezier) is a B-spline using Bernstein polynomials for weighting the influence of the control points (knots).

4.3.3.3 Surfaces

GM_Polygon

A polygon (class GM_Polygon) is a surface patch that is defined by a set of boundary curves and an underlying surface to which these curves adhere. The default is that the curves are coplanar and the polygon uses planar interpolation in its interior.

GM_Triangle

A triangle (class GM_Triangle) is a planar polygon defined by 3 corners.

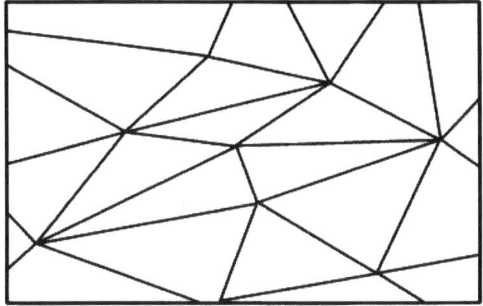

Fig. 4.13. GM_Triangle

GM_ParametricCurveSurface

The surface patches that make up the parametric curve surfaces are all continuous families of curves. A parametric curve surface can be expressed by a mathematical function.

GM_GriddedSurface

The gridded surface (class GM_GriddedSurface) is a GM_ParametricCurveSurface defined from a rectangular grid in the parameter space. The rows from this grid are control points for horizontal surface curves; the columns are control points for vertical surface curves.

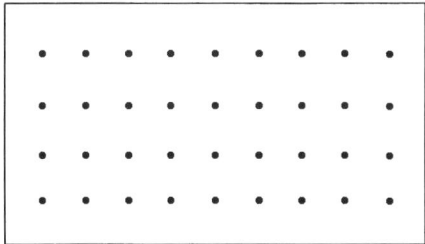

Fig. 4.14. GM_GriddedSurface

GM_Cone

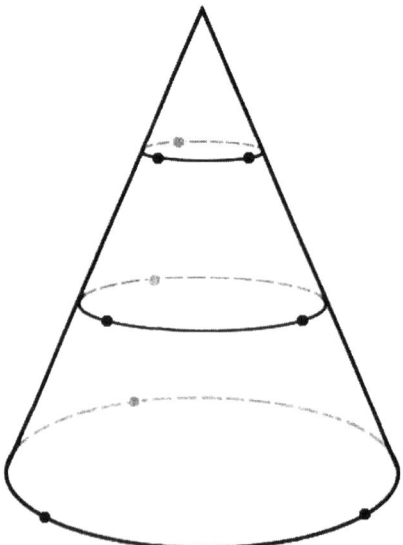

Fig. 4.15. GM_Cone

A cone (class GM_Cone) is a gridded surface given as a family of conic sections whose controlPoints vary linearly. That means that the control points defining the cone in figure 4.15 lie on three straight lines.

GM_Cylinder

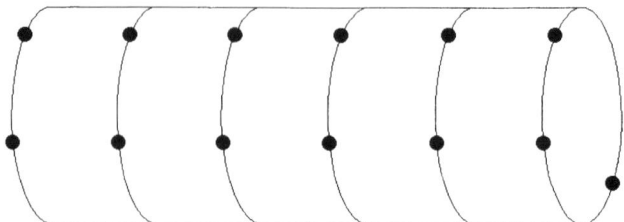

Fig. 4.16. GM_Cylinder

A cylinder (class GM_Cylinder) is a gridded surface given as a family of circles whose positions vary along a set of parallel lines, keeping the cross sectional horizontal curves of a constant shape.

GM_Sphere

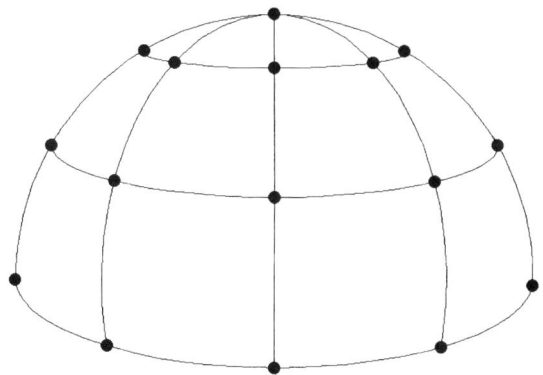

Fig. 4.17. GM_Sphere

A sphere (class GM_Sphere) is a gridded surface given as a family of circles whose positions vary linearly along the axis of the sphere, and whose radius varies in proportion to the cosine function of the central angle. The horizontal circles resemble lines of constant latitude, and the vertical arcs resemble lines of constant longitude.

GM_BilinearGrid

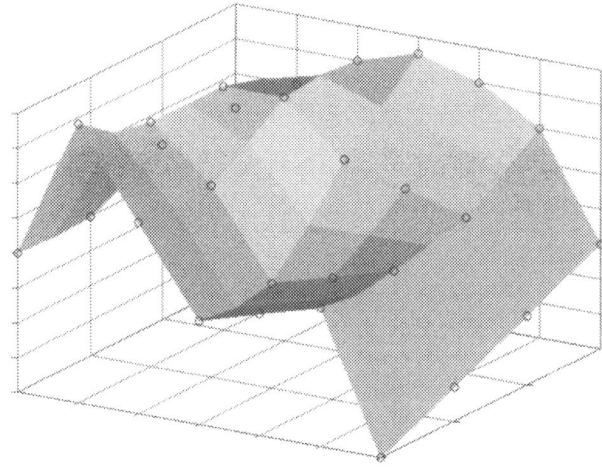

Fig. 4.18. Bilinear grid (Brozio 2003)

A bilinear grid (class GM_BilinearGrid) is a gridded surface that uses line strings as the horizontal and vertical curves.

GM_BicubicGrid

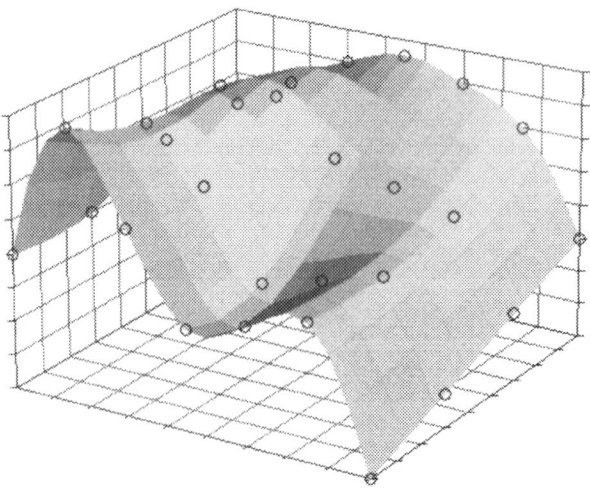

Fig. 4.19. Bicubic grid (Brozio 2003)

A bicubic grid (class GM_BicubicGrid) is a gridded surface that uses cubic polynomial splines as the horizontal and vertical curves.

GM_BSplineSurface

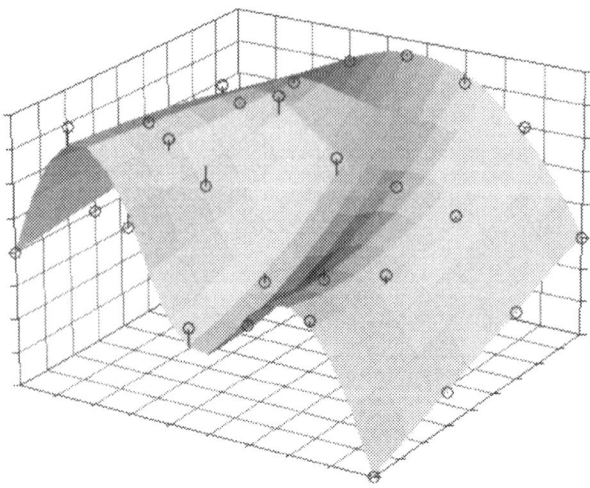

Fig. 4.20. B-spline surface (Brozio 2003)

A B-spline surface (class GM_BSplineSurface) is a rational or polynomial parametric surface that is represented by control points, basis functions and possibly weights. If the weights are all equal then the spline is piecewise polynomial. If they are not equal then the spline is piecewise rational.

GM_PolyhedralSurface

A polyhedral surface (class GM_PolyhedralSurface) is a surface composed of polygon surfaces (class GM_Polygon) connected along their common boundary curves. This differs from the generic surface (class GM_Surface) only in the restriction on the types of surface patches acceptable.

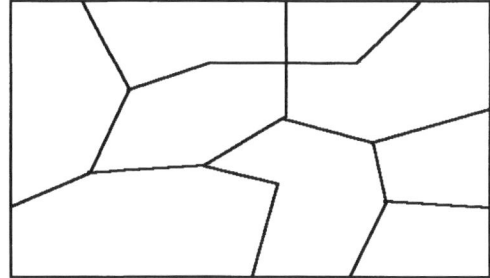

Fig. 4.21. GM_PolyhedralSurface

GM_TriangulatedSurface

A triangulated surface (class GM_TriangulatedSurface) is a polyhedral surface that is composed only of triangles (class GM_Triangle). There is no restriction on how the triangulation is derived.

GM_Tin

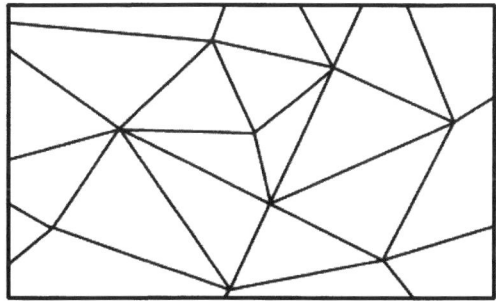

Fig. 4.22. GM_Tin

A TIN (class GM_Tin) is a triangulated surface that uses the Delaunay algorithm or a similar algorithm complemented with consideration for breaklines, stoplines, and maximum length of triangle sides. The Delaunay criterion means that the circumsphere of any of the triangles must not include other points of the point set. The Delaunay triangulation method is commonly used to produce TIN tesselations with triangles that are optimally equiangular in shape. The length of their edges is a minimum. Thiessen polygons are most commonly used in order to optimize the search for the Delaunay criterion.

4.3.3.4 Operations

mbRegion

The operation "mbRegion" returns a region in the Coordinate Reference System that contains this GM_Object. The most common use of mbRegion will be to support indexing methods that use extents other than minimum bounding rectangles (MBR or envelopes).

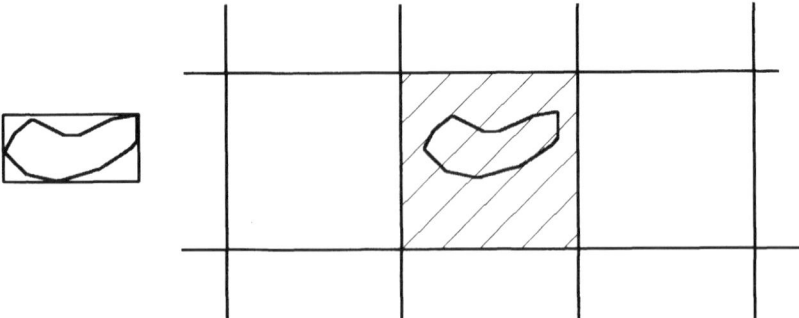

Fig. 4.23. Operation mbRegion; left: minimum bounding rectangle, right: region

representativePoint

The operation "representativePoint" returns a point value (DirectPosition) that is guaranteed to be on this GM_Object.

4.3 Spatial schema (ISO 19107)

Fig. 4.24. Operation representativePoint

boundary

Applied to an object the operation "boundary" returns the object's boundary that is one dimension lower than the dimension of the object itself. In the case of a surface the boundary operator returns the curves around the surface. The object may have a further internal structure.

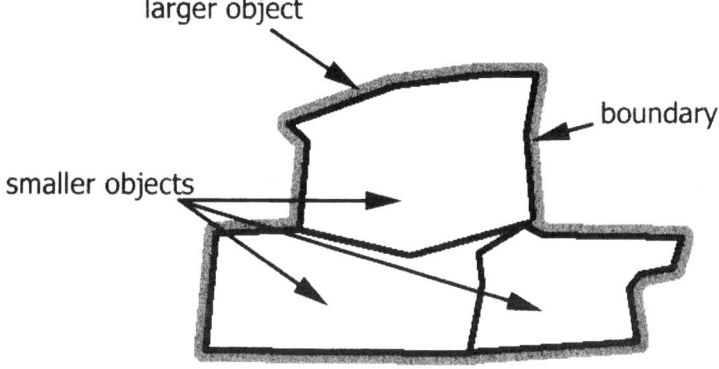

Fig. 4.25. Operation boundary

closure

The operation "closure" combines an object with its boundary. Applied to a GM_LineString the operation "closure" returns the GM_LineString and its start and end point.

isSimple

The operation "isSimple" shall return TRUE if this GM_Object has no interior point of self-intersection or self-tangency.

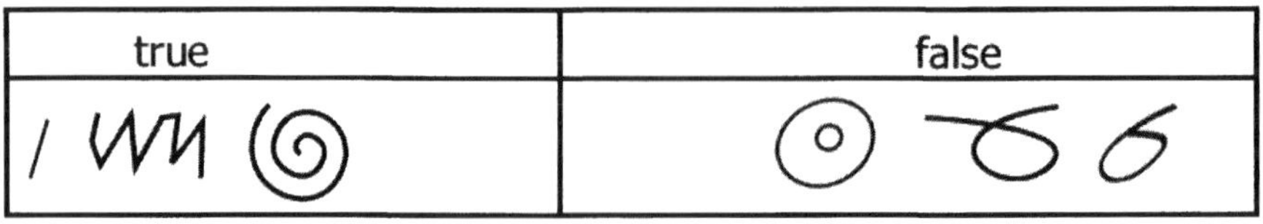

true	false
/ ∧∧ ⊚	⊙ ⌀ ⌀

Fig. 4.26. Operation isSimple

isCycle

The operation "isCycle" shall return TRUE if this GM_Object has an empty boundary after topological simplification. A closed curve or a sphere have empty boundaries. This condition is alternatively referred to as being "closed" as in a "closed curve". This creates some confusion since there are two distinct and incompatible definitions for the word "closed". The use of the word cycle is rarer but leads to less confusion. Essentially, an object is a cycle if it is isomorphic to a geometric object that is the boundary of a region in some Euclidean space. Thus a curve is a cycle if it is isomorphic to a circle (has the same start and end point). A surface is a cycle if it is isomorphic to the surface of a sphere, or some torus.

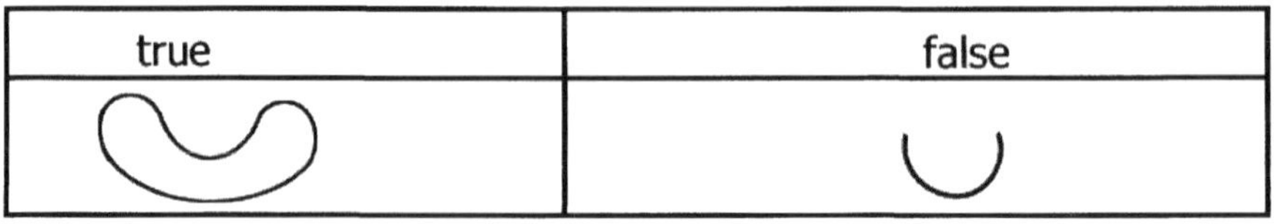

true	false
⌒	⌣

Fig. 4.27. Operation isCycle

distance

The operation "distance" shall return the distance between this GM_Object and another GM_Object. This distance is defined to be the greatest lower bound of the set of distances between all pairs of points that include one each from each of the two GM_Objects.

Fig. 4.28. Operation distance

transform

The operation "transform" shall return a new GM_Object that is the coordinate transformation of this GM_Object into the passed Coordinate Reference System within the accuracy of the transformation.

Fig. 4.29. Operation transform

envelope

The operation "envelope" returns the minimum bounding box for this GM_Object. The minimum bounding rectangle is the 2-dimensional case. This is the coordinate region spanning the minimum and maximum value for each ordinate taken on by DirectPositions in this GM_Object. The simplest representation for an envelope consists of two DirectPositions, the first one containing all the minimums for each ordinate and second one containing all the maximums.

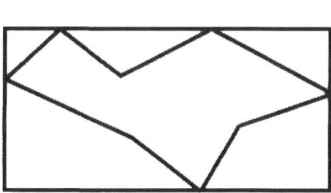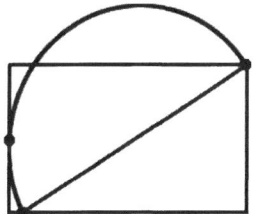

Fig. 4.30. Operation envelope

centroid

The operation "centroid" shall return the mathematical centroid for this GM_Object. The result is not guaranteed to be on the object. For heterogeneous collections of primitives, the centroid only takes into account those of the largest dimension. For example, when calculating the centroid of surfaces, an average is taken weighted by area. Since curves have no area they do not contribute to the average.

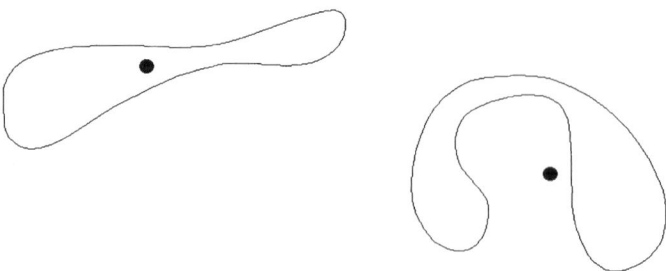

Fig. 4.31. Operation centroid

convexHull

The operation "convexHull" shall return a GM_Object that represents the convex hull of this GM_Object.

Fig. 4.32. Operation convexHull

buffer

The operation "buffer" shall return a GM_Object containing all points whose distance from this GM_Object is less than or equal to the "distance" passed as a parameter.

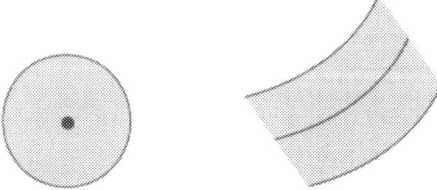

Fig. 4.33. Operation buffer

contains

The Boolean valued operation "contains" returns TRUE if this GM_Object contains another GM_Object, or a single point given by a coordinate (DirectPosition).

Fig. 4.34. Operation contains

intersects

The Boolean valued operation "intersects" returns TRUE if this GM_Object intersects another GM_Object.

Fig. 4.35. Operation intersects

equals

The Boolean valued operation "equals" returns TRUE if this GM_Object is equal to another GM_Object.

Fig. 4.36. Operation equals
Both objects on the left are congruent.

intersection

The "intersection" operation returns the set theoretic intersection of this GM_Object and the passed GM_Object.

Fig. 4.37. Intersection

union

The "union" operation shall return the set theoretic union of this GM_Object and the passed GM_Object.

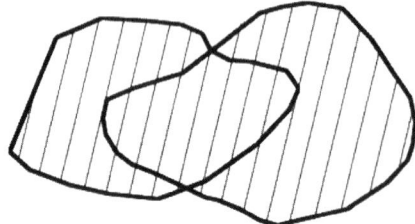

Fig. 4.38. Union

difference

The "difference" operation returns the set theoretic difference of this GM_Object and the passed GM_Object.

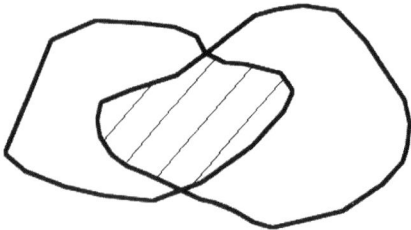

Fig. 4.39. Difference

symmetricDifference

The "symmetricDifference" operation returns the set theoretic symmetricDifference of this GM_Object and the passed GM_Object.

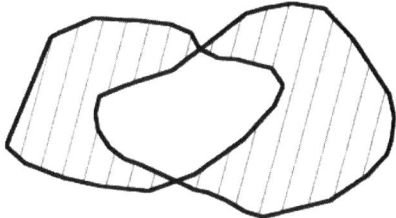

Fig. 4.40. Symmetric Difference

bearing

The operation "bearing" returns the bearing, as a unit vector, of the tangent (at this GM_Point) to the curve between this GM_Point and a passed GM_Position.

area

The operation "area" shall return the sum of the surface areas of all of the boundary components of a solid.

volume

The operation "volume" shall return the volume of this solid (class GM_Solid). Holes in the volume do not count for the volume.

startPoint, endPoint

The operations "startPoint" and "endPoint" return the direct positions of the first point and the last point respectively on a generic curve. This differs from the boundary operator in primitives (class GM_Primitive), since it returns only the values of these two points, not representative objects.

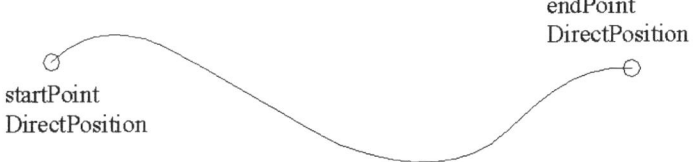

Fig. 4.41. Operations startPoint, endPoint

tangent

The operation "tangent" shall return the tangent vector along this generic curve (class GM_GenericCurve) at the passed parameter value. This vector approximates the derivative of the parameterization of the curve. The tangent shall be a unit vector (have length 1.0).

length

The operation "length" returns the distance between the two points along the curve.

samplePoint

The operation "samplePoint" returns an ordered point array (class GM_PointArray) that lie on the curve segment. In most cases, these will be related to control points used in the construction of the segment.

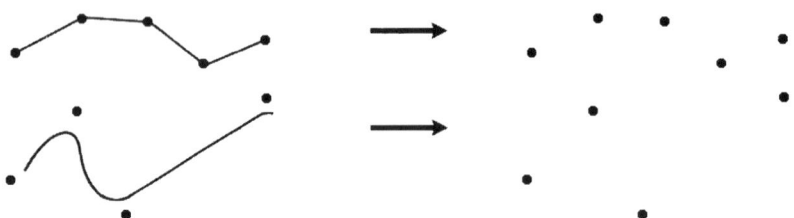

Fig. 4.42. Operation samplePoint

asGM_LineSegment

The operation asGM_LineSegment decomposes a line string into an equivalent sequence of line segments.

Fig. 4.43. Operation asGM_LineSegment

asGM_Geodesic

The operation "asGM_Geodesic" decomposes a geodesic string into an equivalent sequence of geodesic segments.

Fig. 4.44. Operation asGM_Geodesic

center

The operation center calculates the center of the circle of which this arc has a direct position. The Coordinate Reference System of the returned DirectPosition will be the same as that for the arc.

Fig. 4.45. Operation center

radius

The operation radius calculates the radius of the circle of which this arc is a portion.

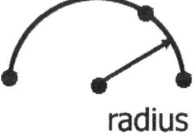

Fig. 4.46. Operation radius

startOfArc

The operation startOfArc calculates the bearing of the line from the center of the circle of which this arc is a portion to the start point of the arc. In the 2-dimensional case this will be a start angle. In the 3-dimensional case, the normal bearing angle implies that the arc is parallel to the reference circle. If this is not the case, then the bearing must include altitude information.

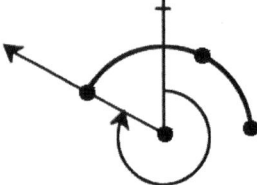

Fig. 4.47. Operation startOfArc

endOfArc

The operation endOfArc calculates the bearing of the line from the center of the circle of which this arc is a portion to the end point of the arc. In the 2-dimensional case this will be an end angle. In the 3-dimensional case, the normal bearing angle implies that the arc is parallel to the reference circle. If this is not the case, then the bearing must include altitude information.

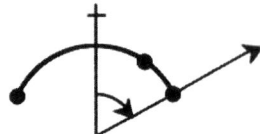

Fig. 4.48. Operation endOfArc

upNormal

The operation "upNormal" returns a vector perpendicular to the generic surface (class GM_GenericSurface) at the DirectPosition passed, which must be on the generic surface.

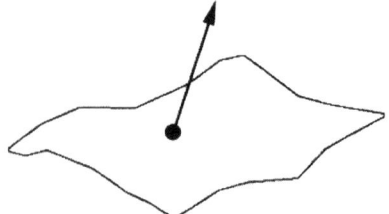

Fig. 4.49. Operation upNormal

perimeter

The operation "perimeter" shall return the sum of the lengths of all the boundary components of this generic surface. Since perimeter, like length, is an accumulation of distance, its return value shall be in a reference system appropriate for measuring distances.

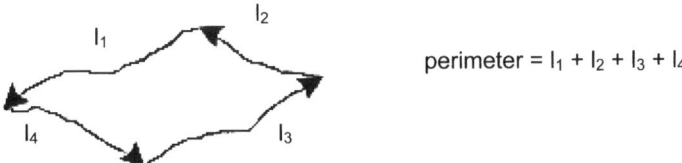

$$\text{perimeter} = l_1 + l_2 + l_3 + l_4$$

Fig. 4.50. Operation perimeter

area

The area of a 2-dimensional geometric object is a numeric measure of its surface area (in a square unit of distance). The operation "area" returns the area of the requested generic surface.

horizontalCurve

The operation "horizontalCurve" constructs a curve that traverses the surface horizontally.

verticalCurve

The operation "verticalCurve" constructs a curve that traverses the surface vertically.

surface

The operation "surface" traverses the surface both vertically and horizontally.

4.3.3.5 Aggregation

Arbitrary aggregations of geometric objects are possible. These are not assumed to have any additional internal structure. The aggregation of points is called multi point, multi curve, multi surfaces, and multi solids (classes GM_MultiPoint, GM_MultiCurve, MultiSurface, and GM_MultiSolid respectively).

4.3.4 Topology

Topology describes the neighbourhood relations of geometric data and is mostly used to accelerate computational geometry. Topology is an abstraction of the underlying geometry. Points where two or more curves meet are called nodes. The curves between pairs of nodes are geometrically simplified to straight lines and called edges. The surfaces surrounded by edges are called faces. The term to describe 3-dimensional bodies defined by nodes, edges and faces is a topological solid.

The root class of topology is TP_Object. It has the subclasses TP_Primitive and TP_Complex. The class TP_Primitive is specialized to its subclasses TP_Node, TP_Edge, TP_Face, and TP_Solid.

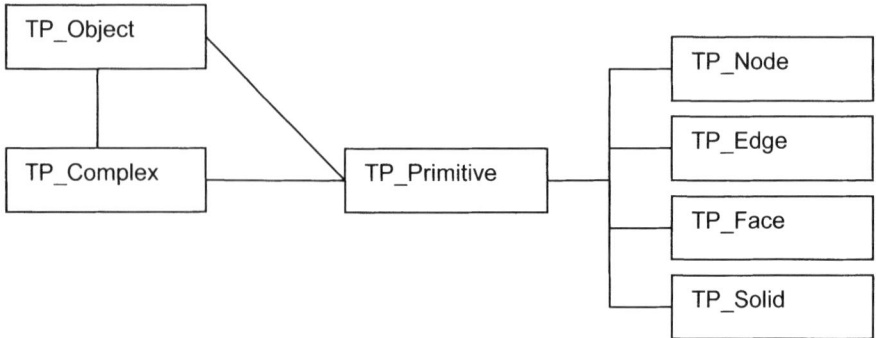

Fig. 4.51. Hierarchy of TP_Object

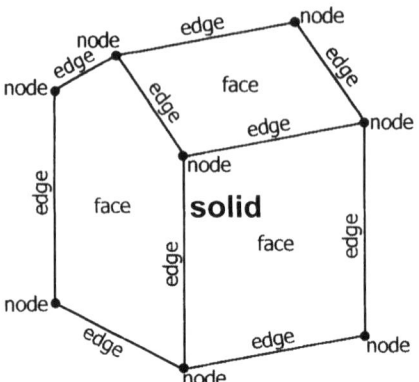

Fig. 4.52. Topology of TP_Object

A topological complex (class TP_Complex) is a complete network of topological elements (node, edge, face, solid). A topological complex is used to describe that two or more topological networks are disjunct. If they are disjunct then more than one topological complex exists.

Two topological complexes may be overlayed without being linked. This may occur if a dataset has two or more thematic layers with different topologies.

4.4 Simple Features (ISO 19125-1)

The ISO 19125-1 covers 2-dimensional geometries with linear interpolation between vertices. The simple features model consists of the root class Geometry and its subclasses Point, Curve, Surface, and GeometryCollection. The class GeometryCollection has the subclasses MultiPoint, MultiCurve, and MultiSurface.

The class Geometry is the equivalent to the class GM_Object in ISO 19107. The class GeometryCollection is the equivalent to the class GM_Aggregate in ISO 19107. The model of ISO 19125-1 does not include complexes, a third dimension, non-linear curves, or topology.

The simple features gained some acceptance in web mapping applications because the amount of required data is much less than in the case of ISO 19107. It is expected that in the future, sophisticated applications like cadastre and cartography will tend to use the ISO 19107 and the ISO 19136 (Geography Markup Language).

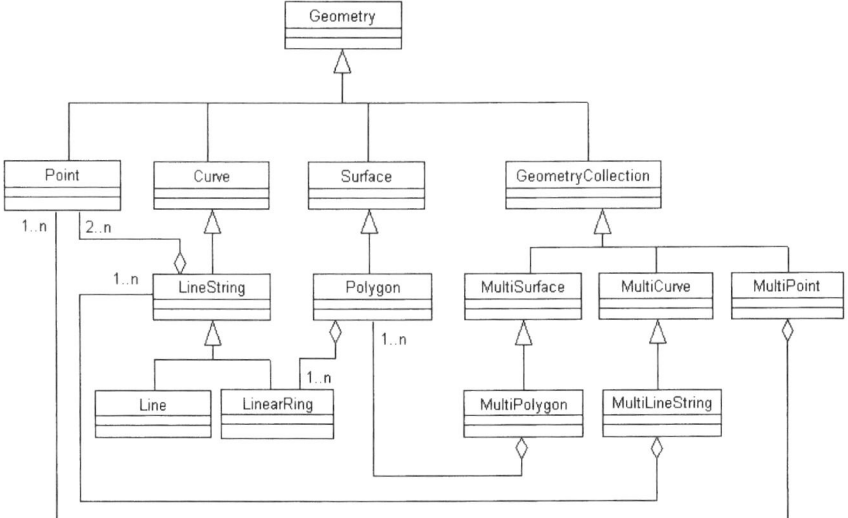

Fig. 4.53. Hierarchy of geometry of Simple Features (ISO54 2000)

4.4.1 Point and MultiPoint

A point (class Point) is a 0-dimensional geometric object and represents a single location in the coordinate space. A multi point (class MultiPoint) is a 0-dimensional geometry collection (class GeometryCollection). The elements of a multi point are restricted to points.

4.4.2 Curve and MultiCurve

A curve (class Curve) is single a line string or a collection of line strings (class LineString).
A line string has a linear interpolation between points (class Point).
A line (class Line) is a line string with exactly 2 points.
A linear ring (class LinearRing) is a line string that is both closed and simple.
The curve in figure 4.54 – (3) is a closed line string that is a linear ring. The curve in figure 4.54 – (4) is a closed line string that is not a linear ring.

John Rowley is from the United Kingdom. He is a Chartered Engineer holding a Bachelor of Science in Civil Engineering. Until 1980, his major pre-occupation was the design of Bridges and the development of associated computer methods with the status of Principal Engineer in his company. In 1987 he established his own GIS consultancy, GEOBASE Consultants Limited. In the last decade the major fields of work have been national, European and international consultancy and standardisation.

During his career Mr. Rowley was encouraged to establish a department that would develop the entire flow line from digital survey methods to the computer aided design for Highways and Bridges. This involved software development as well as procurement. Successful projects were developed for the UK, Libya, Kuwait, Saudi Arabia, and others. In particular, new kinds of projects evolved, especially multi-disciplinary town planning projects involving GIS methods. Among the numerous projects his consultancy company carried a strategic study into the application of GIS in Police methods for the UK Home Office or participated in several studies aimed at improving the Spatial Data Infrastructure in Europe.

Mr. Rowley's several functions in standardisation include the Head of UK delegation from 1992-2001 and a direct involvement in standards such as ISO 19125, ISO 19128. He was Convener of ISO/TC211 WG 8 dealing with Location Based Services and the chairman of the former TC211/OGC Coordination Group that has been replaced by a new structure today.

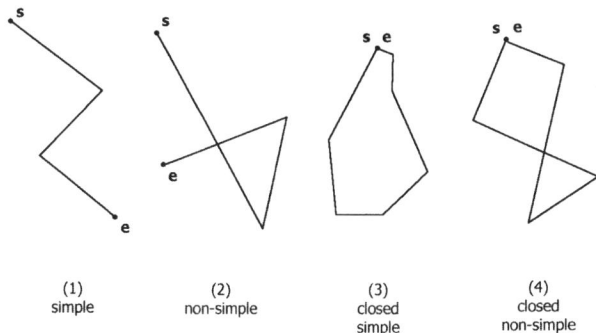

Fig. 4.54. LineString (ISO54 2000)

MultiCurve

A multi curve (class MultiCurve) is a 1-dimensional geometry collection (class GeometryCollection) whose elements are curves as in figure 4.55.

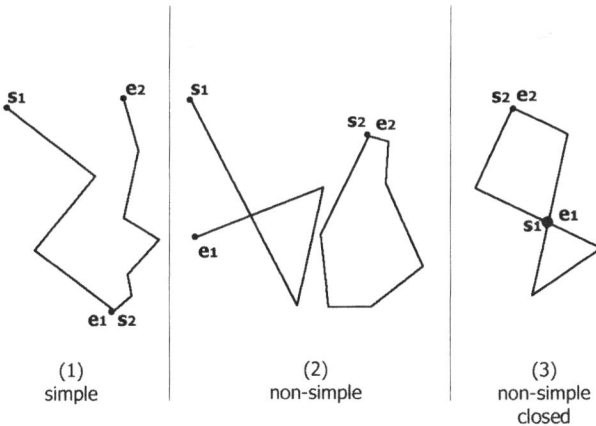

Fig. 4.55. MultiLineString (ISO54 2000)

The multi curve is defined by the following most important characteristics:

A multi curve is simple if and only if all of its elements are simple and the only intersections between any two elements occur at points that are on the boundaries of both elements.

A multi curve is closed if all of its elements are closed. The boundary of a closed multi curve is always empty.

4.4.3 Polygon and MultiPolygon

A surface (class Surface) consists of one or many polygons (class Polygon).

A polygon (class Polygon) is a planar area defined by one exterior boundary and zero or more interior boundaries. Each interior boundary defines a hole in the polygon.

The most important rules that define valid polygons are as follows:
- Polygons are topologically closed.
- The boundary of a polygon consists of a set of linear rings (class LinearRing) that make up its exterior and interior boundaries.
- The interior of every polygon is a connected point set. This is the area within the polygon.
- The exterior of a polygon with one or more holes is not connected. Each hole defines a connected component of the exterior.

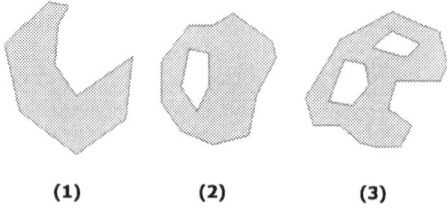

Fig. 4.56. Polygons with 1, 2, and 3 LinearRings respectively (ISO54 2000)

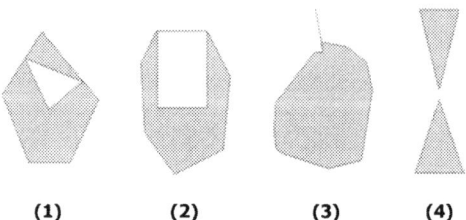

Fig. 4.57. Examples of objects not representable as a single instance of Polygon (ISO54 2000)

MultiPolygon

A multi polygon (class MultiPolygon) is a multi surface (class MultiSurface) whose elements are polygons.

The most important rules for multi polygons are as follows:
- The interiors of 2 polygons that are elements of a multi polygon may not intersect.
- The boundaries of any 2 polygons that are elements of a multi polygon may not "cross" and may touch at only a finite number of Points.

(Worboys 1992; Worboys and Bofakos 1993; Clementini and Di Felice 1995; Clementini and Di Felice 1996)

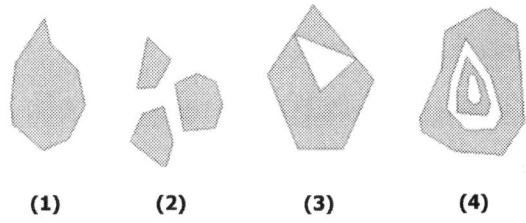

Fig. 4.58. Examples for MultiPolygon (ISO54 2000)

4.4.4 Example for the topology in ISO 19107 and ISO 19125-1

The following diagram (figure 4.59) explains the influence of underlying topology to a geometric dataset.

The top-example displays the map as it would appear on a screen or a print. It contains roads, houses, parcel boundaries and trees. In order to focus on the features and their relation, the graphic portrayal is kept simple. Only the trees are drawn as symbols for clarification.

The lower examples display the same map being exploded to reveal the underlying relations. The lower left graphic shows the geometry and the topology as standardised in ISO 19107. The two lower right graphics show the geometry as standardised in ISO 19125-1.

ISO 19107: Topology relates neighbouring features. This is mostly achieved by not allowing more than one point at a given position. Consequently, two or more neighbouring features that have points at the same position, share the coordinates. Internally the coordinates of the points are stored only once while the features relate their positions to this point using a pointer technique. The topology can be used to accelerate geometric computations such as a network-search, or to track gaps and overlaps in the dataset.

ISO 19125-1: As this standard does not support topology, the features are geometrically linked by a common Coordinate Reference System only. A full set of coordinates is stored together with every feature. This means that at one position, more than one point can exist. This approach is well known from CAD-type systems.

Smaller features within larger surfaces require cut-outs. As cut-outs are not supported by ISO 19125-1, features like houses or trees have to sit on the highest display level meaning that they have to be plotted last. This is the upper case. Alternatively cut-outs can be simulated by introducing a cut in the surface enabling a continuous sequence of points on a perimeter that includes the perimeter of the cut-out.

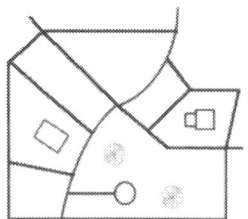

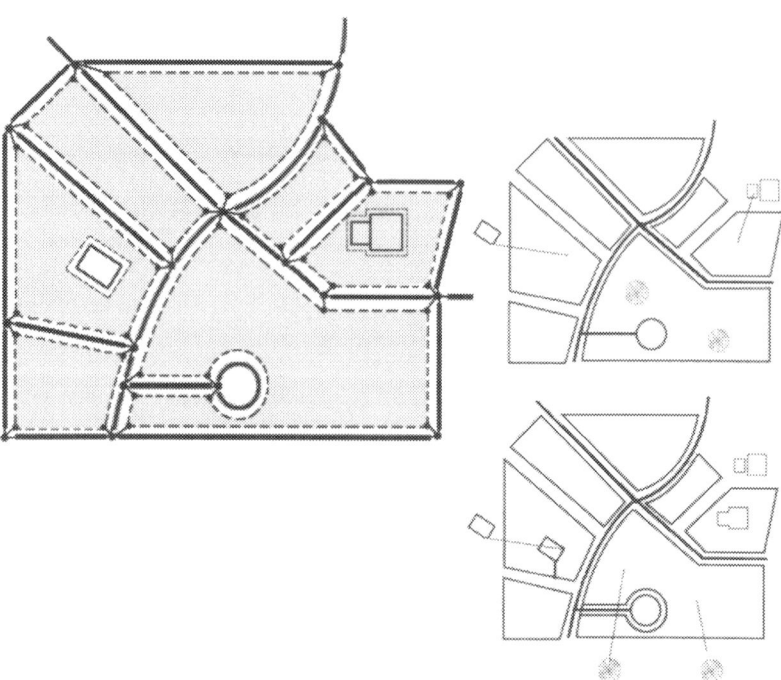

Fig. 4.59. Neighbourhood relations in Spatial schema and Simple Features.

4.5 Schema for coverage geometry and functions (ISO 19123)

Coverages are geographic information schemas that integrate discrete and continuous geographic phenomena. Discrete phenomena are recognizable objects that have relatively well-defined boundaries or spatial extent. Examples would include buildings, streams, and measurement stations. Continuous phenomena vary over space and have no specific extent. Examples of this would include temperature, soil

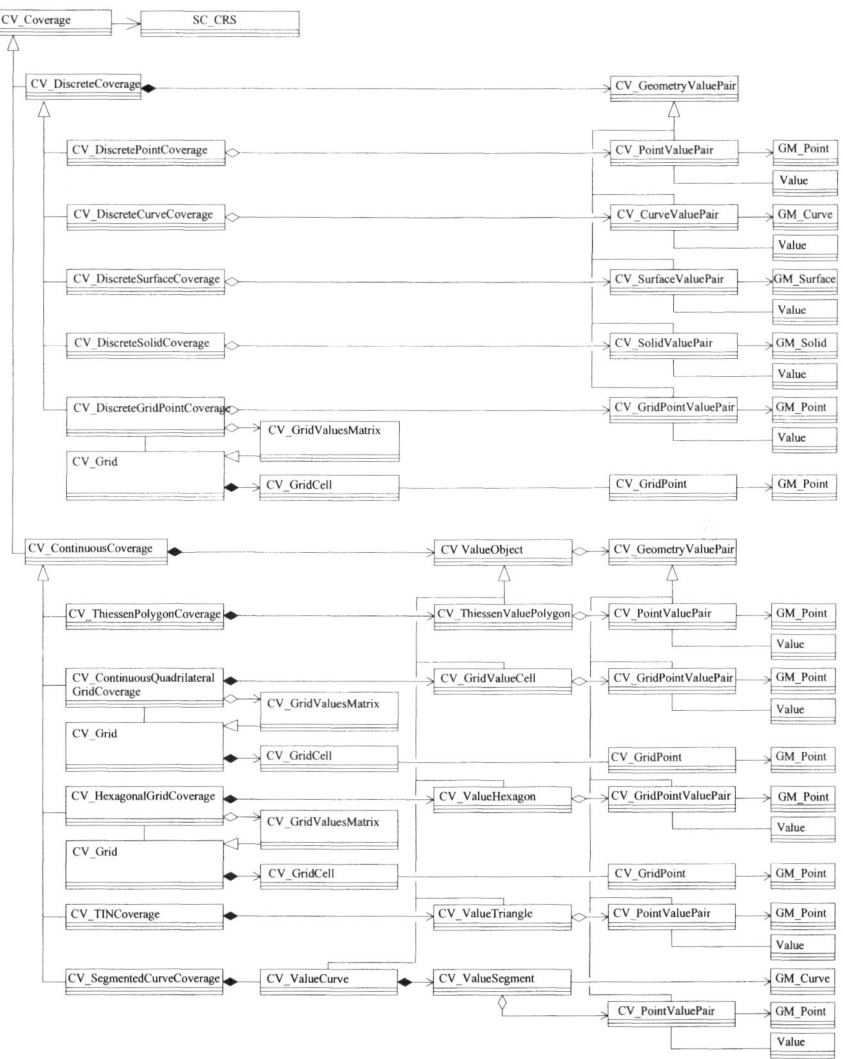

Fig. 4.60. Hierarchy of CV_Coverage (ISO 19123)

composition, and elevation. A value or description of a continuous phenomenon is only meaningful at a particular position in space and possibly time. Temperature, for example, takes on specific values only at defined locations whether measured or interpolated from other locations.

4.5.1 Overview over ISO 19123

A coverage is a subtype of a geographic feature (class GF_FeatureType, ISO 19109) and is associated to a Coordinate Reference System according to ISO 19111.

4.5.2 Description of the discrete coverage classes of ISO 19123

A discrete coverage consists of individual geometric objects in which an interpolation between those objects is not allowed.

CV_DiscretePointCoverage

A discrete point coverage (class CV_DiscretePointCoverage) is a set of irregularly distributed points and their attributes. The principle use of discrete point coverages is to provide a basis for continuous coverage functions.

An example for a discrete point coverage would be a small scale map showing cities and their population. In this case it is not feasible to interpolate between the points of the cities.

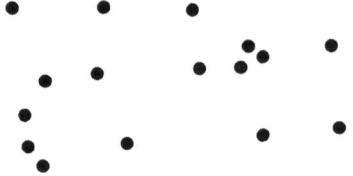

Fig. 4.61. Discrete point coverage

CV_DiscreteCurveCoverage

A discrete curve (class CV_DiscreteCurveCoverage) coverage is characterized by a finite spatiotemporal domain consisting of curves that may be elements of a network.

An example of a discrete curve coverage is a road-network where the coverage assigns a route number, a name, a pavement width, and a pavement material type to each segment of the road system.

4.5 Schema for coverage geometry and functions (ISO 19123)

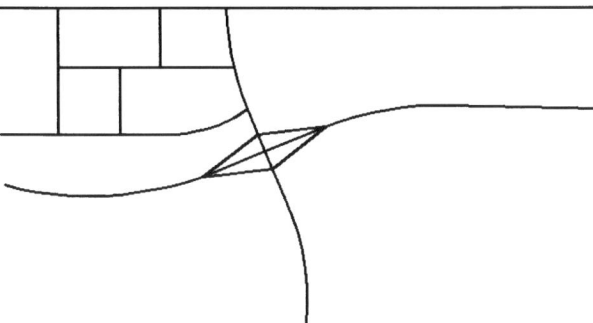

Fig. 4.62. Discrete curve coverage

CV_DiscreteSurfaceCoverage

A discrete surface coverage (class CV_DiscreteSurfaceCoverage) consists of a collection of surfaces. In most cases, the surfaces that constitute the spatiotemporal domain of a coverage are mutually exclusive and exhaustively partition the extent of the coverage

For example, a coverage that represents corn fields typically has a spatial domain composed of a surface with irregular boundaries where no interpolation is feasible.

Fig. 4.63. Discrete surface coverage

CV_DiscreteSolidCoverage

A discrete solid coverage (class CV_DiscreteSolidCoverage) is a coverage in which its domain consists of a collection of solids.

For example, an ocean body or an atmospheric volume could be represented as a CV_DiscreteSolidCoverage with a range of attributes such as temperature and pressure at each vertex.

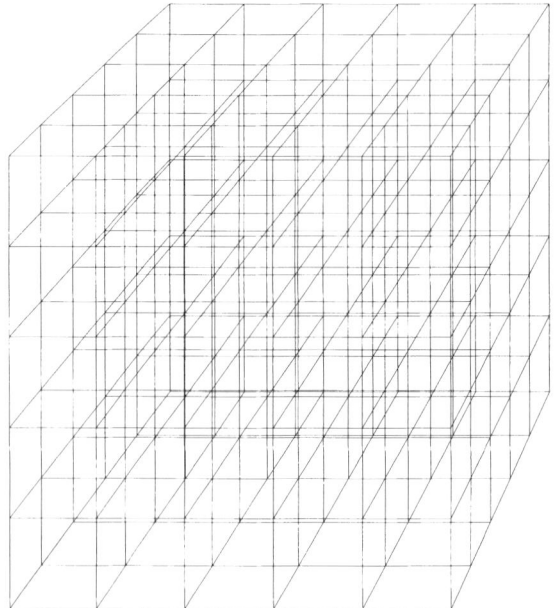

Fig. 4.64. Discrete solid coverage

The 3D-grid could be a volume such as an ocean body. Every vertex may hold a range of attributes.

CV_DiscreteGridPointCoverage

A discrete grid point coverage (class CV_DiscreteGridPointCoverage) is a set of regularly distributed points and their attributes. The distribution is defined in the grid values matrix (class CV_GridValuesMatrix). A discrete grid point coverage could be considered a subcase of the discrete point coverage (class CV_DiscretePointCoverage).

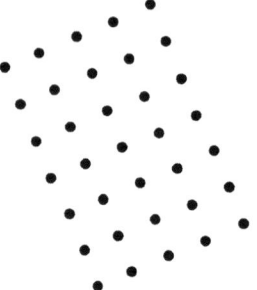

Fig. 4.65. Discrete grid point coverage

4.5.3 Description of the continuous coverage classes of ISO 19123

A continuous coverage is a combination of a discrete coverage with an interpolation function that fills the space and eventually the time in between. For the interpolation the ISO 19123 defines the following methods:

Table 4.1. Relation between interpolation methods and coverage types

Interpolation method	Type of coverage that uses this interpolation method
Nearest neighbour interpolation	any coverage
Linear interpolation	segmented curve coverage
Quadratic interpolation	segmented curve coverage
Cubic interpolation	segmented curve coverage
Bilinear interpolation	quadrilateral grid coverage
Biquadratic interpolation	quadrilateral grid coverage
Bicubic interpolation	quadrilateral grid coverage
Lost area interpolation	discrete point coverages
Barycentric interpolation	TIN coverage

CV_ThiessenPolygonCoverage

A Thiessen polygon coverage (CV_ThiessenPolygonCoverage) partitions the space using Thiessen polygons. Thiessen polygons are closed polygons around points of an irregularly distributed point set. For any position within a Thiessen polygon its centre point is the closest of the whole point set. Thiessen polygons are defined in the 2-dimensional space only. They make it possible to apply lost area interpolation to a point set. They are also used for one step in the computation of a TIN for a given point set.

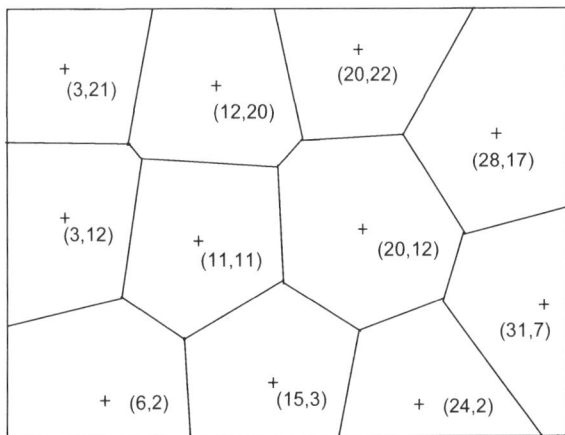

Fig. 4.66. Thiessen polygon network (ISO53 2002)

CV_ContinuousQuadrilateralGridCoverage

The quadrilateral grid coverage (class CV_ContinuousQuadrilateralGridCoverage) employs a systematic tessellation of the spatiotemporal domain. The principle advantage of such tessellations is that they support a sequential enumeration of elements that makes data storage and access more efficient.

A quadrilateral grid coverage may be a rectified grid or a referenceable grid. A rectified grid can be transformed into a Coordinate Reference System using an affine transformation. An affine transformation has six parameters: two translations (x, y), two rotations (one for each axis), and two scales (one for each axis). A referenceable grid requires a formula with a higher order that transforms into a Coordinate Reference System. An example is the perspective transformation with 8 parameters.

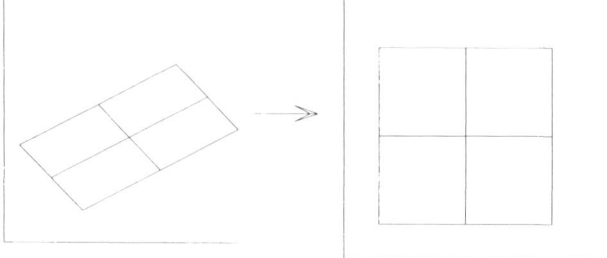

Fig. 4.67. Rectified grid (affine transformation)

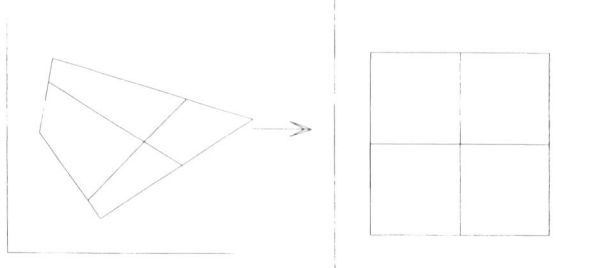

Fig. 4.68. Referenceable grid (non-affine transformation)

CV_HexagonalGridCoverage

A hexagonal grid coverage (class CV_HexagonalGridCoverage) is based on regular hexagons. That grid can be described as a rectified grid in which the two offset vectors are of equal length but differ in direction by 60°.

4.5 Schema for coverage geometry and functions (ISO 19123)

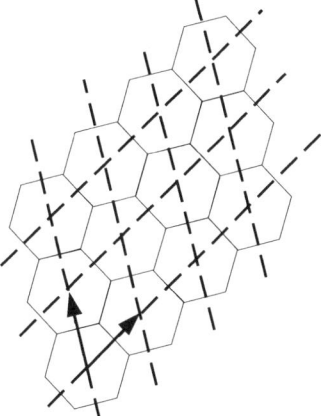

Fig. 4.69. Hexagonal grid coverage (ISO53 2002)

CV_TINCoverage

A TIN coverage (class CV_TINCoverage) partitions the space in triangles that are formed according to the Delaunay criterion (see chapter 4, section 3.3.3, GM_Tin).

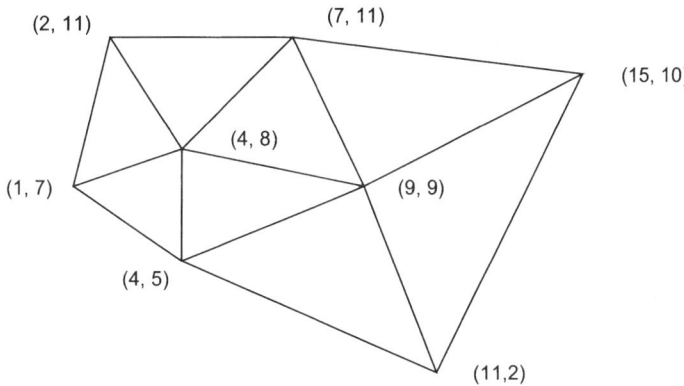

Fig. 4.70. Triangulated irregular network (Tin) (ISO53 2002)

CV_SegmentedCurveCoverage

The segmented curve coverages (class CV_SegmentedCurveCoverage) are used to model phenomena that vary continuously or discontinuously along curves that may be elements of a network. An example is a road network with variable traffic intensity along the roads.

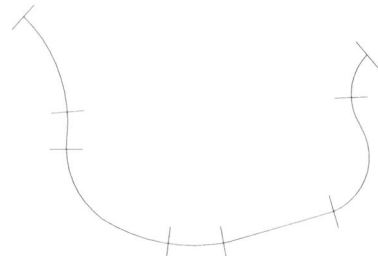

Fig. 4.71. Segmented curve coverage

CV_DomainObject

A domain object (class CV_DomainObject) represents an element of the spatiotemporal domain of the coverage. It is an aggregation of objects that may include any combination of GM_Objects or other spatial or temporal objects defined in the ISO 19100 standards. A domain object is used to address spatial or temporal parts of the coverage like the grid points or larger portions.

CV_ValueObject

A value object (class CV_ValueObject) represents an element of the spatiotemporal domain of a continuous coverage. A value object is used to address spatial or temporal parts of the continuous coverage like a triangle of the TIN. A value object may be the result of an evaluation of a continuous coverage and thus data not persistent in the coverage.

CV_GeometryValuePair

A geometry value pair (class CV_GeometryValuePair) describes an element of a coverage. As the name implies, each geometry value pair consists of two parts: a geometric object like a point and a record of associated feature values like its elevation and its soil type.

CV_GridPointValuePair

A grid point value pair (class CV_GridPointValuePair) is composed of a grid point and a feature attribute value record.

4.5 Schema for coverage geometry and functions (ISO 19123)

CV_GridValuesMatrix

The grid values matrix (class CV_GridValuesMatrix) basically defines the attributive values at the grid points and the geometric sequence of their occurrence within the grid. For example, the grid values matrix of 100 x 100 elevation points contains the 10,000 elevation values and a pointer to the sequence rule like combewise from upper left to lower right.

The ISO 19123 standardises six sequence rules. The graphic examples relate all to a 2-dimenional grid.

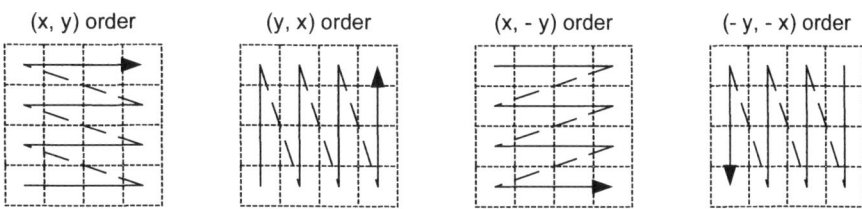

Fig. 4.72. Linear scanning (ISO53 2002)

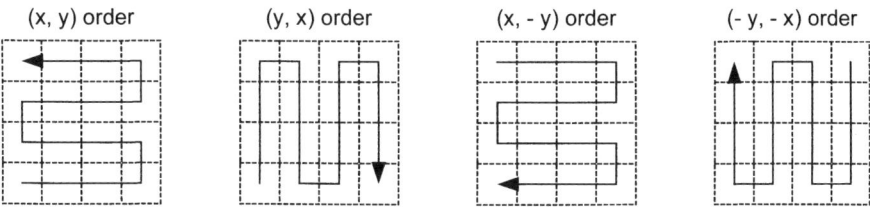

Fig. 4.73. Boustrophedonic scanning (ISO53 2002)

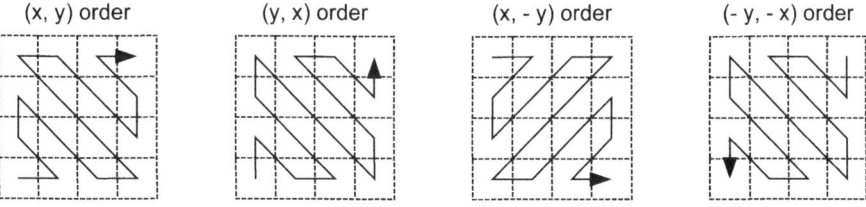

Fig. 4.74. Cantor-diagonal scanning (ISO53 2002)

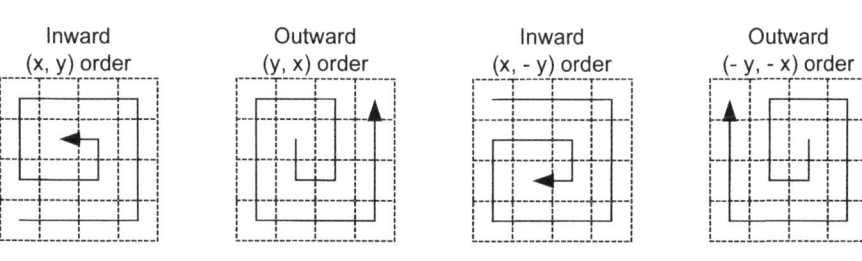

Fig. 4.75. Spiral scanning (ISO53 2002)

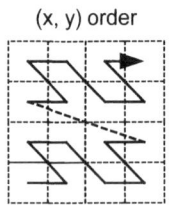 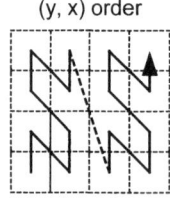 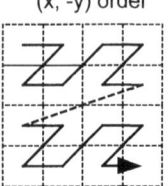 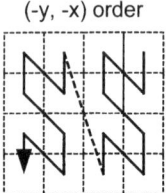

Fig. 4.76. Morton order (ISO53 2002)

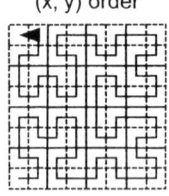 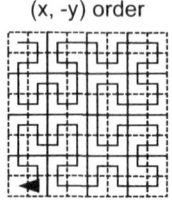 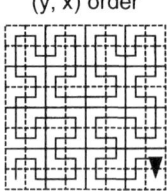 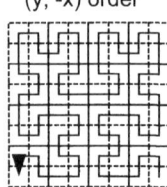

Fig. 4.77. Hilbert order (ISO53 2002)

4.5.4 Operations in ISO 19123

Table 4.2. Operations in ISO 19123

	evaluate	evaluate Inverse	locate, (curve, segment)	find	list
CV_Coverage	x	x			
CV_DiscreteCoverage	x	x	x	x	x
CV_DiscretePointCoverage	x	x	x	x	x
CV_DiscreteCurveCoverage	x	x	x	x	x
CV_DiscreteSurfaceCoverage	x	x	x	x	x
CV_DiscreteSolidCoverage	x	x	x	x	x
CV_DiscreteGridPointCoverage	x	x	x	x	x
CV_ContinuousCoverage	x	x	x		
CV_ThiessenPolygonCoverage	x	x	x		
CV_ContinuousQuadrilateralGridCoverage	x	x	x		
CV_HexagonalGridCoverage	x	x	x		
CV_TINCoverage	x	x	x		
CV_SegmentedCurveCoverage			x		

4.5 Schema for coverage geometry and functions (ISO 19123)

The table 4.2 relates the important operations of coverages to the classes in ISO 19123. Some operations of the subclasses of CV_TINCoverage and CV_SegmentedCurveCoverage are not shown.

The operations "curve" and "segment" are only used with CV_SegmentedCurveCoverages. All other coverage types use the operation "locate".

evaluation, evaluateInverse, and locate

The operation "evaluate" accepts a position as input and returns a record of attribute values for that position. Most evaluation methods involve interpolation within the neighbourhood of the position.

The operation "evaluateInverse" accepts a record of feature attribute values as input and returns a set of spatiotemporal objects. For example, this operation could return a set of contour lines derived from the feature attribute values associated with the grid points.

The operation "locate" accepts a position as input and returns the set of value objects such as Thiessen polygons that contain this position. It shall return a null value if the position is not in any of the value objects within the spatiotemporal domain of the discrete coverage.

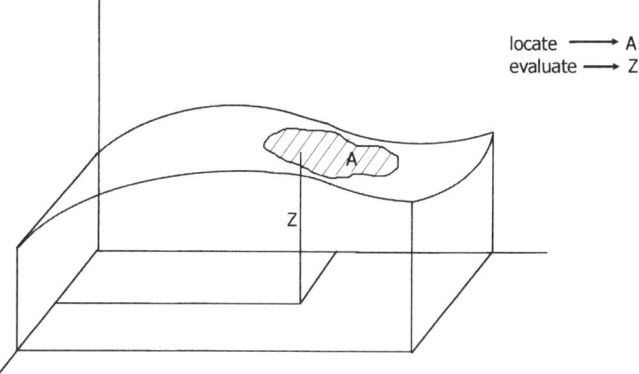

Fig. 4.78. Operations evaluate and locate

In detail, the operation "evaluate" works differently depending on the four different kinds of coverages. Figure 4.79 shows the four cases based on a quadrilateral grid (square grid).

Cell structures and the operation evaluate

The interstitial spaces between grid lines are called grid cells. Grid points viewed as lying on the cell corners are often called posts and data composed of post points and associated feature attributes values is called matrix data. Grid points viewed as lying at the centres of the cells are called cell centres. Users of digital images sometimes take the point of view where it is assumed that each pixel value is the weighted average measurement of the scene, taken over a grid cell centred on a grid point.

There are four cases for the evaluation of the grid:

1. The grid points are considered to be the *post points in a discrete point* coverage. The coverage can only be evaluated at direct positions that fall on the grid points.
2. The grid points are considered to be at the *cell centres in a discrete surface* coverage. In order to evaluate the coverage at some direct position, the nearest grid points have to be identified.
3. The grid points are considered to be *post points in a continuous* grid coverage. The feature attribute values at any direct position are interpolated from those at a set of surrounding grid points.
4. The grid points are considered to be *cell centres in a continuous* grid coverage. The feature attribute values at any direct position are interpolated from those at a set of surrounding grid points.

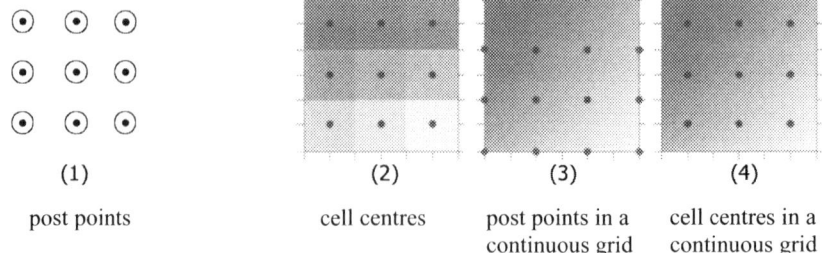

(1) post points (2) cell centres (3) post points in a continuous grid (4) cell centres in a continuous grid

Fig. 4.79. Evaluation of a grid

The common point rule (class CV_CommonPointRule) provides a strategy for the operation "evaluate" to supply an unequivocal return value for all positions of the coverage. The common point rule behaves differently with discrete and continuous coverages. In discrete coverages, an interpolation between the post points is not allowed. In continuous coverages, ambiguities may arise at positions that fall either on a boundary between geometric objects or within the boundaries of two or more overlapping geometric objects. In this case, the interpolation within each geometric object takes place first. The common point rule is then applied.

The common point rule offers four selection techniques: average value, lowest value, highest value, or all range values of the neighbouring post points. In the case of a segmented curve coverage, the startValue or the endValue can also be selected for return.

interpolate

The operation "interpolate" accepts a position as input and returns the record of feature attribute values computed for that position.

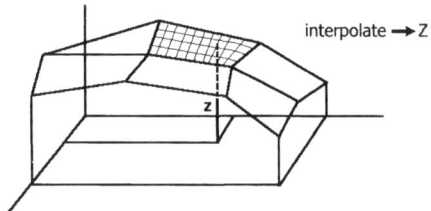

Fig. 4.80. Operation interpolate

Find, list, and curve

The operations "find" and "list" only exist for discrete coverages.

The operation "find" accepts a position as input and returns the sequence of geometry value pairs that include the domain object nearest to the position and their distances from the position.

The operation "list" returns the dictionary that contains the domain objects, like curves of the discrete coverage, each paired with its record of feature attribute values.

The operation "curve" accepts a direct position as input and returns the value curve (class CV_ValueCurve) nearest to that position.

4.6 Geography Markup Language (GML) (ISO 19136)

The ISO 19136 (Geography Markup Language) standardises an implementation of the geometry-related standards of the 19100 family, in particular the ISO 19107 (Spatial schema) and the ISO 19123 (Schema for coverage geometry and functions). However, the ISO 19123 seems to deal more with interfaces while the coverages in GML are described more from an information viewpoint (Portele 2003). The Geography Markup Language (GML) is an application of the Extensible Markup Language (XML) built on XML Schema. This book refers to GML version 3.0.

GML is designed for the modelling, the transport and storage of geographic data. In a number of predefined schemas GML provides a rich vocabulary that can be used to create domain-specific GML application schemas. GML serves as a foundation for the geospatial web and for the interoperability of independently developed distributed applications including location based services.

178   4 Geometry standards

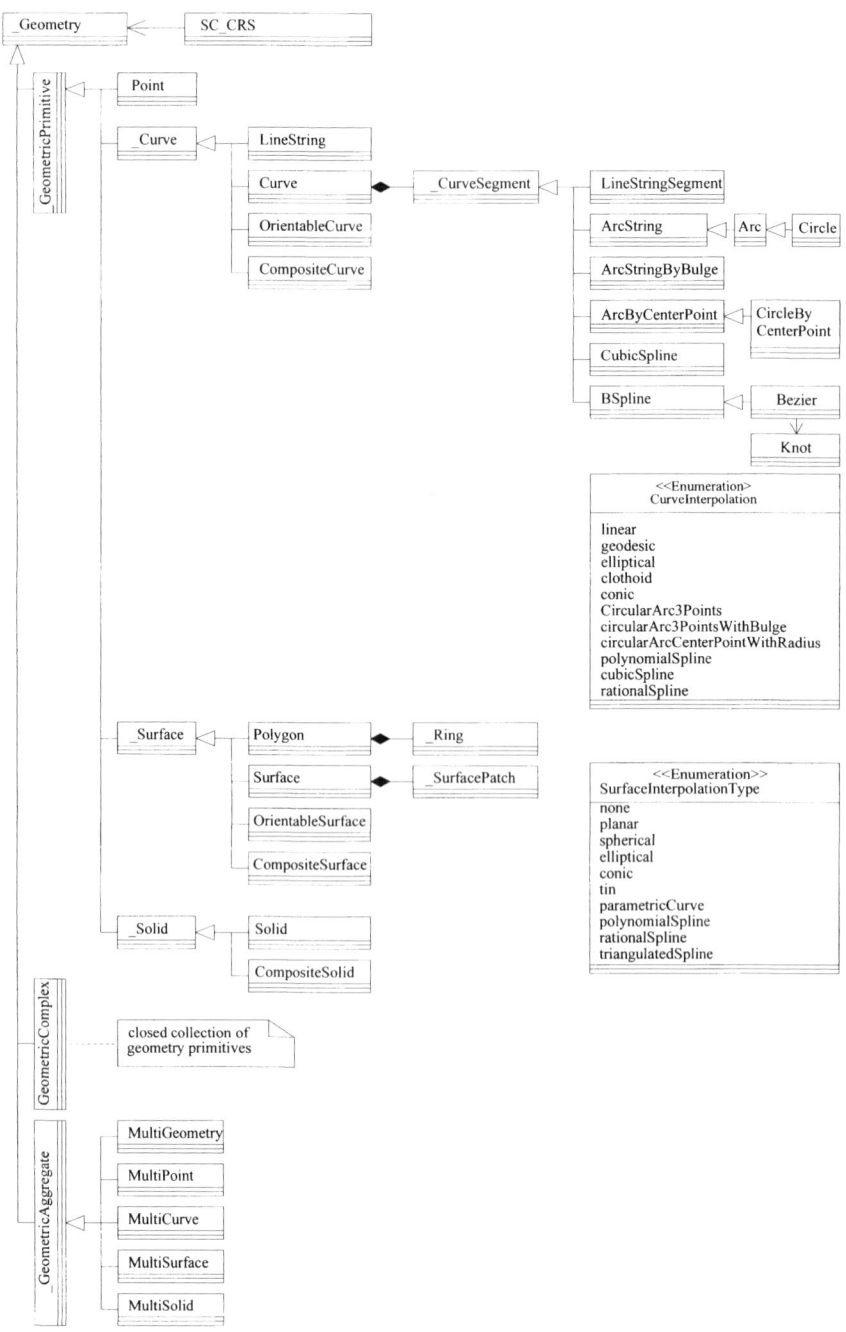

Fig. 4.81. Hierarchy of elements in ISO 19136 (GML 3.0)

4.6.1 GML schemas

Feature and feature collection

A feature is an abstraction of a real world phenomenon. It is a *geographic* feature if it is associated with a location relative to the earth. The state of a feature is defined by a set of properties, where each property can be thought of as a {name, type, value} triple.

A feature collection is a collection of features that can itself be regarded as a feature. As a consequence a feature collection has a feature type and thus may have distinct properties of its own, in addition to the features it contains.

Geographic features in GML include coverages and observation as subtypes.

Geometry

The geometry of a geographic feature describes its location, shape or extent. The geometry model of GML distinguishes geometric primitives, aggregates, and complexes.

The geometric primitives are the basic elements that are used to form the geometry of a geographic dataset. Primitives are open, that is, a curve does not contain its end points, a surface does not contain its boundary curves, and a solid does not contain its bounding surface.

The geometric aggregates are arbitrary aggregations of geometry elements. They are not assumed to have any additional internal structure and are used to "collect" pieces of geometry of a specified type.

Geometric complexes are closed collections of geometric primitives. That means that they contain their boundaries.

Figure 4.81 shows the hierarchy of the GML geometry types.

Coordinate Reference System

The Coordinate Reference System (CRS) provides the meaning for location coordinates. A CRS may be associated with any geometry of GML.

The CRS schema of GML deviates slightly from the ISO 19111 (Spatial referencing by coordinates) and from the Topic 2 (Spatial Referencing by Coordinates) of the OGC's Abstract Specification. Appropriate change proposals will be submitted to the ISO/TC211 (Portele 2003).

Topology

Topology describes the geometric properties of objects that are invariant under continuous deformation. For example a square is topologically equivalent to a rectangle and a trapezoid. In geographic modelling, the foremost use of topology is in accelerating computational geometry.

The topology of a dataset is described by the topological primitives nodes, edges, faces, and topological solids. Nodes are topological points where edges meet. Edges are topological lines where faces meet. Faces are topological surfaces where solids meet.

Topological relations are described with boundaries, coboundaries, and directed topological primitives. The topological primitive "edge" is bounded by two directed topological primitives "node". The topological primitive "edge" is also the coboundary to a pair of nodes. The other relations are formed according to table 4.3.

Table 4.3. Relation between topological primitive, directed primitive, and coboundary

topological primitive --> coboundary <--	dimension	bounded by directed primitive topological primitive	dimension
edge	1	node	0
face	2	edge	1
topological solid	3	face	2

Temporal information and dynamic features

The time in GML allows the description of the time dependent status of geographic features and the description of dynamic features, for example in the domain of location based services. The definitions of time in GML extend the model of ISO 19108 (Temporal schema).

Time is measured in two types of scales: interval and ordinal. An interval scale offers a basis for measuring duration. An ordinal scale provides information only about relative position in time, for example a stratigraphic sequence or the geological time scale.

The default temporal reference system is the Gregorian calendar with UTC (ISO17 2000).

A time instant represents a position in time. It is the equivalent to a point in space. Inexact or "fuzzy" positions may be expressed using the indeterminatePosition attribute that may have values such as "after" or "before".

A period represents an extent in time. It is the equivalent to a curve in space. It is an interval bounded by beginning and end instants, and has a duration.

The status of dynamic features can be described by a snapshot and by a time slice. A snapshot portrays the status of a feature as a whole whereas a time slice encapsulates the dynamic properties that reflect some change event.

Coverage

Coverages in GML are defined in accordance with ISO 19123 (Schema for coverage geometry and functions) and OGC (OGC00 2000). However, GML implements only a subset of the functionality defined in the quoted sources.

Styling

The GML requires a strict separation of data and presentation. Therefore none of the GML data description constructs such as features and geometries have built-in capabilities to describe the styling information. In order to simplify the handling of GML a default styling mechanism was created as a separate model that can be "plugged-in" to a GML data set. The default style schema depends on the W3C Synchronized Multimedia Integration Language (SMIL) (SMIL 2001).

The GML styling mechanism distinguishes between four kinds of styling types. The feature style applies to each feature independently. The geometry style describes the style for one geometry of a feature. The topology style describes the style for one topology property. The label style describes the style for the text that is to be displayed along with the graphical representation.

The presentation consists of styling elements. The symbol element specifies a graphical symbol. It may be defined within the GML application schema. This case is called "inline". It also may be addressed by a pointer to other GML- or non-GML-sources. This case is called "remote".

The style element is the term for simple symbols.

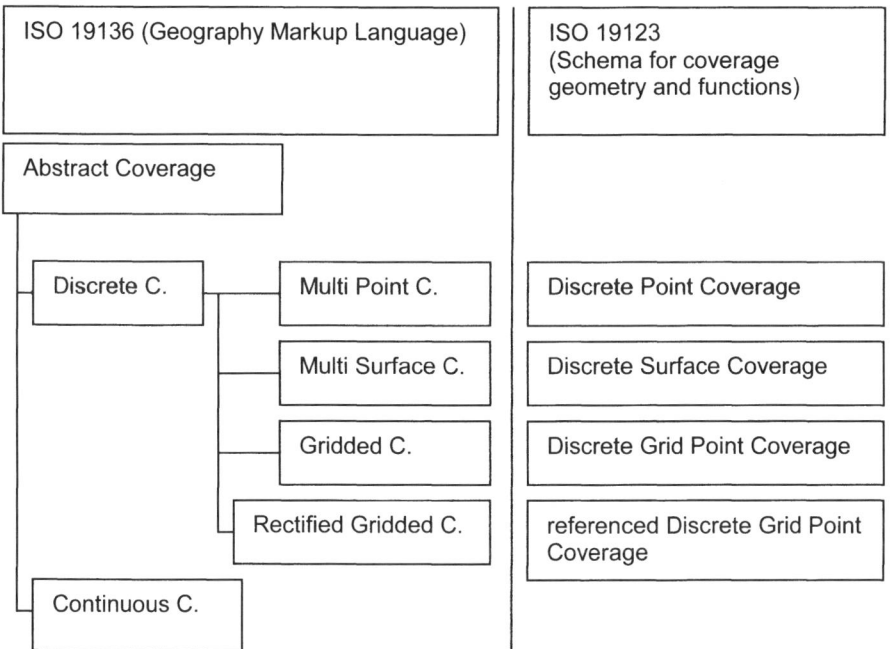

Fig. 4.82. Comparison between the coverage types in ISO 19136 and ISO 19123

The style variation element is used in cases where a symbol is displayed repeatedly, but with slight variations each time. In this case the geometry of the symbol has to be transmitted only once.

Animation attributes are used to describe the animation behaviour of the geometry, topology, label, or graph (SMIL 2001).

XLink

XLink is an XML-technique that explicitly relates two or more data objects or portions of data objects. XLink is widely used in GML. XLink components are used in GML to implement associations between objects by reference. It is currently restricted to unidirectional links between two resources in most GML applications. The associated objects may be physically as close as two elements in the same GML document or as far as any node in the world wide web. Further details related to XLink are described in annex C "Extensible Markup Language (XML) ".

Profile

A profile of GML defines a subset of the constructs of GML. GML profiles use a "copy-and-delete-approach". To create a profile, a developer might copy the applicable schema files from GML and simply delete any global types, elements and local optional particles that he or she does not need for his or her application schema.

A profile can be the basis of an application schema. The building of an application schema is a two-part process. The profile acts as a restriction of GML to produce types and elements consistent with the complete GML 3 but potentially lacking in some optional particles. The application schema then uses these types as a common base, and uses them in new types and elements by extensions or inclusions.

GML 3 --- (selection and restriction) --> GML profile --- (extension and inclusion) --> application schema

The ISO 19136 defines a set of rules that have to be applied for creating a profile. They can be summarized as follows:

1. A profile of GML is a logical restriction of a subset of GML.
2. A profile must not change the name, definition, or data type of mandatory GML elements or attributes.
3. The relevant schema or schemas that define a profile must use the core GML namespace.
4. An application schema may extend and use types from the profile, but must do so in its own namespace.

Other topics of GML 3

In addition to the constructs explained above, GML offers further techniques for the description of data elements:
1. Description of observations
2. Units, measures, and values

3. Names for directions
4. Definitions and dictionaries

4.6.2 GML application schema

A GML application schema is an XML Schema, conforming to the rules outlined in the ISO 19136 document. The application schema is a combination of a GML profile and application-specific definitions. It describes one or more types of geographic objects, components of geographic objects, or meta data, including dictionaries and definitions, used in the definition of geographic objects.

The definition of a GML application schema must follow a number of rules. They shall guarantee the readability of and the ability to process a GML document by any GML parser and shall allow for the worldwide transport and storage of GML datasets without interference with other GML datasets.

The rules have mostly three components:

1. The application schema must declare a target namespace. This is the namespace in which the terms like the application specific feature types and coverage "live". This must not be the GML namespace. The GML namespace is defined in about 20 XML schemas such as geometryPrimitives.xsd and coverage.xsd.

2. All geographic features in the application schema must be declared as global elements in the schema. The content model for such global elements must derive either directly or indirectly from gml:AbstractFeatureType.

3. Elements representing properties of features may be declared as global elements in an application schema, or they may be declared locally within feature content models (type definitions). The type for a property element may be derived from gml:AssociationType or gml:ArrayAssociationType, or may follow the pattern of gml:AssociationType.

4.6.3 Relation between GML, OGC, and the ISO 19100 standards

GML is being developed by the GML Revision Working Group of the Open GIS Consortium (OGC). GML is related to the OGC Abstract Specification and published as one of their Implementation Specifications.

ISO has adopted GML as the ISO 19136 (Geography Markup Language) standard. The standardisation process is ongoing.

Relation to ISO 19107 (Spatial schema)

The implementation standard ISO 19136 conforms to the other standards of the ISO 19100 family. The closest relation exists to the ISO 19107.

The vast majority of concepts in ISO 19107 have been implemented in GML 3.0 such as the fundamental geometries point, curve, surface, and solid, and the description of positions related to a Coordinate Reference System.

However, the conformance with other ISO 19100 standards does not require a 1-to-1 implementation. The implementation of classes in GML is driven by applications and requirements. In fact ISO 19107 states that is not expected that an implementation will implement all classes. On the other hand GML 3 contains some elements that are beyond the range of the other ISO 19100 standards. The class GM_OffsetCurve is an example of a class in ISO 19107 that has not been implemented by GML 3. The element ArcByCenterPoint is an example of an element in GML 3 that does not exist in ISO 19107.

Relation to other ISO 19100 standards

The ISO 19118 sets the encoding for the geomatics standards. Although XML is strongly preferred, the present version (ISO50 2002) allows different encoding schemas. It is planned to make an XML encoding a normative part of ISO 19118 (Encoding) (Annex C) as soon as GML has become an International Standard.

The ISO 19111 (Spatial referencing by coordinates) in general serves the needs of GML 3. Some modifications have been identified and will be fed back to ISO/TC211.

GML 3 contains a mechanism to include metadata, but there is no rule that only ISO 19115 (Metadata) or later ISO 19139 (Metadata – implementation specification) can be used. GML is open to integrate metadata from other source such as the Dublin Core as well.

The ISO 19108 (Temporal schema), the ISO 19109 (Rules for application schema) and the ISO 19117 (Portrayal) still need to be fully harmonized with GML 3.

4.6.4 Alignment between ISO 19136 and GML-development

As the ISO 19136 defines an implementation it will change more rapidly than the abstract standards of the ISO 19100 family. It may happen that the version of GML that finally becomes an International Standard may be outdated because the development of GML will go on in parallel to the process of ISO standardisation. Therefore provisions are made to sustain alignment between the ISO 19136 and GML.

If possible, new versions of GML should be backward compatible or provide for an automatic mapping. The development of GML will have to take care that the conformance with the conceptual model of the ISO 19100 standards is maintained. From the perspective of the geographic information community and the marketplace, if there are good reasons for a change, the ISO 19100 conceptual model, the ISO 19118 (Encoding), or the OGC Abstract Specification, which GML is based on, should change.

5 Liaison members of ISO/TC211

5.1 Internal liaison members of ISO/TC211

The ISO/TC211 (Geographic information / Geomatics) was founded in 1994 to standardise geographic information. However, various aspects of geographic information had been published as ISO, IEC, or ITU standards before the ISO/TC211 was founded and other relevant standards are being developed while the work of the ISO/TC211 is ongoing. Formal relations with the responsible committees, known as "internal liaisons" have been established in order to align the existing standards and ongoing activities with the work of ISO/TC211.

The number of internal liaisons varies according to the needs of the standardisation work. The following list has been assembled from *all* Technical Committees and Subcommittees of ISO and IEC, as well as all study groups of ITU. The groups that are presently not formal internal liaison members are marked.

In many cases, only a very specific Working Group or Subcommittee of a Technical Committee will develop an internal liaison-relationship to the ISO/TC211. An example is the ISO/TC82 (Mining). Only the Subcommittee 1 (ISO/TC82 SC1) titled "Geological and Petrographic Symbols" has an internal liaison with ISO/TC211. The Working Groups and the Subcommittees relevant to geographic information are titled accordingly and placed as the first paragraph of each section. In the situations where the names of the Working Groups and Subcommittees do not clearly define the subjects that they cover, a list of topics is provided for clarification. The other paragraphs in a section point to further Working Groups or Subcommittees of the Technical Committee. An example is the ISO/TC204 (Intelligent transport systems) in which the Working Group 5 (Fee and toll collection) (ISO/TC204 WG5) has only a minor interest in geographic information.

The internal liaison committees with ISO/TC211 (Geographic information / Geomatics) are underlined.

As an introduction a listing of all relevant ISO and IEC TCs and SCs as well as the ITU SGs (Study Groups) is given in brief.

ISO/TC12	Quantities, units, symbols, conversion factors
ISO/TC20	Aircraft and space vehicles
ISO/TC23	Tractors and machinery for agriculture and forestry
ISO/TC37	Terminology and other language resources
ISO/TC42	Photography:
ISO/TC46	Information and documentation
ISO/TC82	Mining
ISO/TC130	Graphic Technology
ISO/TC172	Optics and optical instruments
ISO/TC176	Quality management and quality assurance
ISO/TC184	Industrial automation systems and integration
ISO/TC204	Intelligent transport systems
ISO/TC211	Geographic information / Geomatics
IEC/TC1	Terminology
IEC/TC80	Maritime navigation and radio-communication equipment and systems
ISO/IEC JTC1/SC2	Coded character sets
ISO/IEC JTC1/SC7	Software and systems engineering
ISO/IEC JTC1/SC11	Flexible magnetic media for digital data interchange
ISO/IEC JTC1/SC18	Document processing and related communication (disbanded in 1997)
ISO/IEC JTC1/SC22	Programming languages, their environment and systems software interfaces
ISO/IEC JTC1/SC23	Optical disk cartridges for information interchange
ISO/IEC JTC1/SC24	Computer graphics and image processing
ISO/IEC JTC1/SC25	Interconnection of information technology equipment
ISO/IEC JTC1/SC27	IT security techniques
ISO/IEC JTC1/SC29	Coding of audio, picture, multimedia and hypermedia information
ISO/IEC JTC1/SC32	Data management and interchange
ISO/IEC JTC1/SC34	Document description and processing languages
ISO/IEC JTC1/SC35	User interfaces
ITU-T/SG2	Operational aspects of service provision, networks and performance
ITU-T/SG12	End-to-end transmission performance of networks and terminals
ITU-T/SG13	Multi-protocol and IP-based networks and their Internetworking
ITU-T/SG16	Multimedia services, systems and terminals
ITU-T/SG17	Data networks and telecommunication systems

5.1.1 ISO

ISO/TC12: **Quantities, units, symbols, conversion factors**
Working Group relevant to geographic information:
- Space and time (WG4)

ISO/TC20: **Aircraft and space vehicles**
Subcommittee relevant to geographic information:
- Space data and information transfer systems (SC13)

ISO/TC23: **Tractors and machinery for agriculture and forestry**
Subcommittee relevant to geographic information:
- Agricultural electronics (SC19)

ISO/TC37: **Terminology and other language resources**
(presently no internal liaison with ISO/TC211)

ISO/TC42: **Photography:**
(presently no internal liaison with ISO/TC211)
Working Groups relevant to geographic information:
- Electronic still picture imaging (WG18)
- Digital still cameras (JWG20, Joint WG with IEC)
- Colour management (JWG22, Joint WG with IEC/TC100 and ISO/TC130)
- Extended colour encoding for digital image (WG23, joint with ISO/130 and CIE)

Topics not mentioned in the Working Group names:
- Digital storage media
- Film scanners

Other Working Groups and topics:
- Photoflash units (WG2)
- Sensitometry, image measurement and viewing (WG3)
- Photographic chemicals and processing (WG6)
- Still projectors and transparencies (WG9)
- Photographic cameras
- Film (including aerial film)

ISO/TC46: **Information and documentation**
Working Groups relevant to geographic information:
- Coding of country names and related entities (WG2)
- The Dublin core metadata element set

Other topics:
- Transliteration of Arabic and Thai characters into Latin

ISO/TC82: **Mining**
Subcommittee relevant to geographic information:
- Geological and petrographic symbols (SC1)

ISO/TC130: **Graphic Technology**
Working Groups relevant to geographic information:
- Prepress data exchange (WG2)
- Process control and related metrology (WG3)

Topics not mentioned in the Working Group names:
- TIFF/IT (TIFF for image technology)
- Colour
- Proof and production prints
- Displays for colour proofing
- PDF: CMYK, colour managed workflows

Other topics:
- Printing ink
- Register pin systems
- Testing of prints and printing paper

ISO/TC172: **Optics and optical instruments**
(presently no internal liaison with ISO/TC211)
Subcommittees relevant to geographic information:
- Laser (Electro-optical systems) (SC9)
- Geodetic and surveying instruments (SC6)

Other Subcommittees:
- Optical material (SC3)
- Telescopic systems (SC4)
- Microscopes (SC5)

ISO/TC176: **Quality management and quality assurance**

ISO/TC184: **Industrial automation systems and integration**
Subcommittee relevant to geographic information:
- Industrial data (SC4)

Topic not mentioned in the Subcommittee name:
- EXPRESS language

ISO/TC204: **Intelligent transport systems**
Working Groups relevant to geographic information:
- Architecture (WG1)
- Transport Information and Control System (TICS) database technology (WG3)
- General fleet management and commercial/freight (WG7)

- Public transport/emergency (WG8)
- Integrated transport information, management and control (WG9)
- Traveller information systems (WG10)
- Route guidance and navigation systems (WG11)
- Dedicated short range communication for TICS (WG15)
- Wide area communications/protocols and interfaces (WG16)

Topics not mentioned in the Working Group names:
- Geographic Data Files (GDF)
- On board navigation system architecture
- Application Interface Definition for Global Navigation Satellite
- Requirements for interactive centrally determined route guidance
- Intermodal goods transport

Others Working Groups and topics:
- Automatic vehicle and equipment identification (WG4)
- Fee and toll collection (WG5)
- Vehicle/roadway warning and control systems (WG14)
- Systems and Cellular Networks
- Lane departure warning systems
- Manoeuvring aids for low speed operation

ISO/TC211: **Geographic information / Geomatics**

5.1.2 IEC

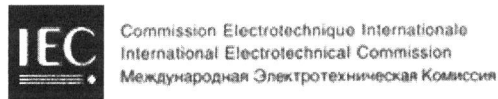

IEC/TC1: **Terminology**
(presently no internal liaison with ISO/TC211)

IEC/TC80: **Maritime navigation and radio-communication equipment and systems**
(presently no internal liaison with ISO/TC211)

Working Groups relevant to geographic information:
- Terrestrial position fixing aids (WG4)
- Global Navigation Satellites Systems (GNSS) (WG4a)
- Electronic Chart Display and Information System (ECDIS) (WG7)

Topics not mentioned in the Working Group names:
- Decca, Omega, Loran C

5.1.3 ISO/IEC JTC1

ISO/IEC JTC1/SC2: Coded character sets

ISO/IEC JTC1/SC7: Software and systems engineering
(presently no internal liaison with ISO/TC211)
> Working Groups relevant to geographic information:
> - Open Distributed Processing (ODP) Enterprise viewpoint (WG17)
> - Modelling languages, metadata, ODP framework and ODP components (WG19)
>
> Topic not mentioned in the Working Group names:
> - Open Distributed Processing – Reference Model (ODP-RM)

ISO/IEC JTC1/SC11: Flexible magnetic media for digital data interchange
(presently no internal liaison with ISO/TC211)

ISO/IEC JTC1/SC18: Document processing and related communication
(disbanded in 1997)
> Working Group relevant to geographic information:
> - Standard Generalized Markup Language (SGML) (WG8)

ISO/IEC JTC1/SC22: Programming languages, their environment and systems software interfaces
(presently no internal liaison with ISO/TC211)
> Working Groups relevant to geographic information:
> - Pascal (WG2) no longer active
> - Fortran (WG5)
> - Basic (WG8) no longer active
> - C (WG14)
> - C++ (WG21)

ISO/IEC JTC1/SC23: Optical disk cartridges for information interchange
(presently no internal liaison with ISO/TC211)

ISO/IEC JTC1/SC24: Computer graphics and image processing
> Working Groups relevant to geographic information:
> - Multimedia Presentation and Interchange (WG6)
> - Image Processing and Interchange (WG7)
> - Environmental Representation (WG8)
>
> Topics not mentioned in the Working Group names:
> - Graphical Kernel System (GKS) (WG6)

- Programmer's Hierarchical Interactive Graphics System (PHIGS) (WG6)
- Portable Network Graphics (PNG) (WG6)
- Basic Image Interchange Format (BIIF) (WG7)

Historical work items:
- GKS-3D
- Computer Graphics Metafile
- Interface techniques for dialogues with graphical devices (CGI)

ISO/IEC JTC1/SC25:   Interconnection of information technology equipment
(presently no internal liaison with ISO/TC211)
Working Group relevant to geographic information:
- Interconnection of computer systems and attached equipment (WG4)

Topics not mentioned in the Working Group names:
- Characteristics of Local Area Networks (LAN)
- Fibre Distributed Data Interface (FDDI)
- Small Computer System Interface (SCSI)

ISO/IEC JTC1/SC27:   IT security techniques
(presently no internal liaison with ISO/TC211)

ISO/IEC JTC1/SC29:   Coding of audio, picture, multimedia and hypermedia information
(presently no internal liaison with ISO/TC211)
Working Groups relevant to geographic information:
- Coding of still pictures (WG1)
- Coding of moving pictures and audio (WG11)

Topic not mentioned in the Working Group names:
- JPEG 2000

ISO/IEC JTC1/SC32:   Data management and interchange
Working Group relevant to geographic information:
- SQL multimedia & application packages (WG4)

ISO/IEC JTC1/SC34:   Document description and processing languages
(presently no internal liaison with ISO/TC211)
Topics not mentioned in the Working Group names:
- Standard Generalized Markup Language (SGML)
- Hypertext Markup Language (HTML)

ISO/IEC JTC1/SC35:   User interfaces
Working Groups relevant to geographic information:
- Keyboards and input interfaces (WG1)
- User interface interaction (WG2)

- Graphical symbols (WG3)
- User interfaces for mobile devices (WG4)
- Cultural, linguistic and user requirements (WG5)
- User interfaces for people with special needs – including the elderly and disabled (WG6)

5.1.4 ITU

The relevant ITU-T Study Groups are mentioned in the brief listing at the beginning of chapter 5, section 1 (presently no internal liaison with ISO/TC211).

5.2 External liaison organizations to ISO/TC211

5.2.1 Portrait of all (status July 2003)

The work of standardisation committees relies heavily on the expertise of organizations outside the usual standardisation business. These organizations, known as "external liaison members", are assigned to the Technical Committees of ISO or IEC according to the needs of a specific standards development project.

Even though the external liaison members of ISO/TC211 are all international representatives of some kind of geographic information, each are a different type of organization with a different organizational structure. They might be categorized in the following way:

International organizations, representing
- Government agencies; example are IHO (hydrography) and WMO (weather)
- Scientific subjects; examples are IAG (geodesy) and FIG (surveying)
- Regional interests; examples are JRC (Europe) and PCGIAP (Asia, Pacific)
- Some tasks of the United Nations; an example is UNGEGN (geographic names)

The following list is a complete summary of all current external liaison members of ISO/TC211. The Internet address is provided for further information.

As an introduction a listing of all external liaison members is given in brief.

CEOS/WGISS   Committee on Earth Observation Satellites/Working Group on Information Systems and Services

DGIWG	Digital Geographic Information Working Group
EPSG	European Petroleum Survey Group
FAO/UN	Food and Agriculture Organization of the United Nations
GSDI	Global Spatial Data Infrastructure
GRSS	IEEE Geoscience and Remote Sensing Society
IAG	International Association of Geodesy
ICA	International Cartographic Association
ICAO	International Civil Aviation Organization
FIG	International Federation of Surveyors
IHB, IHO	International Hydrographic Bureau, International Hydrographic Organization
ISPRS	International Society for Photogrammetry and Remote Sensing
ISCGM	International Steering Committee for Global Mapping
JRC	Joint Research Centre of the European Union
OGC	Open GIS Consortium, Incorporated
PCGIAP	Permanent Committee on GIS Infrastructure for Asia and the Pacific
PCIDEA	Permanent Committee on Spatial Data Infrastructure for the Americas
SCAR	Scientific Committee on Antarctic Research
UN ECE	United Nations Economical Commission for Europe, Statistical Division
UNGIWG	United Nations Geographic Information Working Group
UNGEGN	United Nations Group of Experts on Geographical Names
WMO	World Meteorological Organization

Committee on Earth Observation Satellites/Working Group on Information Systems and Services (CEOS/WGISS)

CEOS is the primary international organization that provides a forum for the co-ordination of the EO (Earth Observation) programmes of world's space agencies together with the primary global environmental programmes including IGOS (International Global Observing Strategy), the Global Observing Systems (GOOS, GCOS, GTOS) and IGBP (International Geosphere Biosphere Programme). WGISS is one of the two standing Working Groups of CEOS (the other being the Working Group on Calibration and Validation). WGISS aims to stimulate, co-ordinate and monitor

EO initiatives, thereby enabling users at global, regional and local level to exploit more effectively and benefit from data generated by EO satellites and other sources.

Scope: In 1995, CEOS established the Working Group on Information Systems and Services (WGISS). It addresses the issues of capture, description, processing, access, retrieval, utilisation, maintenance and exchange of spaceborne EO data and supporting information, enabling improved interoperability and interconnectivity of information systems and services.

Objectives:
- To enable EO data and information services to be more accessible and usable to data providers and data users worldwide through international coordination.
- To enhance the complementarity, interoperability and standardisation of EO data and information management and services.
- To foster the easier exchange of EO data and information through networks and other means.

Chair: Canada
Homepage: http://wgiss.ceos.org
Members: 36 members
Examples are: NASA (USA), ESA (Europe), NASDA (Japan), BNSC (United Kingdom), CNES (France), CSA (Canada), RSA (Russia), DLR (Germany), FAO (UN, food and agriculture), ISPRS (photogrammetry and remote sensing), and UNEP (UN, environment)

Digital Geographic Information Working Group (DGIWG)

Digital Geographic Information Exchange Standard (DIGEST)

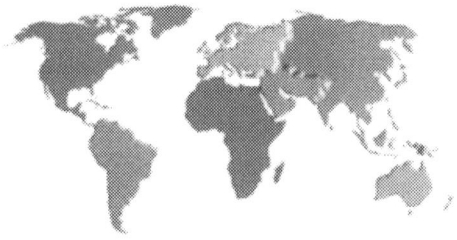

DGIWG represents the experts on digital geographic information in the military agencies primarily of the NATO countries. It maintains the DIGEST geographic data exchange standard for vector data, raster data, and topological information.

Objectives: Established in 1983, the DGIWG developed the Digital Geographic Information Exchange Standard (DIGEST) to support effi-

	cient exchange of digital geographic information among nations, data producers and data users.
Members:	Australia, Belgium, Canada, Czech Republic, Denmark, France, Germany, Greece, Italy, Netherlands, New Zealand, Norway, Portugal, Spain, Turkey, USA, United Kingdom
Chair:	UK
Homepage:	http://www.digest.org/Overview2.htm
See also:	ISO 19120 (Functional standards)

European Petroleum Survey Group (EPSG)

EUROPEAN PETROLEUM SURVEY GROUP

The EPSG is a European group gathering the interests of the oil company surveyors. EPSG has assembled an almost complete dataset of the parameters describing all global coordinate systems being used worldwide.

Objectives:	Founded in 1986, the EPSG maintains and publishes a dataset of parameters for coordinate system and coordinate transformation description. The parameters are maintained in an MS Access relational database and may be downloaded from the homepage. The EPSG geodetic parameters have been included as reference data in the GeoTIFF data exchange specifications and in other international data models.
Meetings:	Twice a year
Chair:	United Kingdom
Homepage:	http://www.epsg.org

Food and Agriculture Organization of the United Nations (FAO/UN)

The Food and Agriculture Organization of the United Nations (FAO) was founded in 1945 with a mandate to raise levels of nutrition and standards of living, to improve agricultural productivity, and to better the condition of rural populations. Today, FAO is one of the largest specialized agencies in the United Nations system and the lead agency for agriculture, forestry, fisheries and rural development.

Objectives:	FAO gives practical help to developing countries through a wide range of technical assistance
	FAO serves as a clearing-house, providing farmers, scientists, government planners, traders and non-governmental organizations with the information they need to make rational decisions on planning, investment, marketing, research and training.
	FAO advises on national strategies for rural development, food security and the alleviation of poverty.
Members:	183 member countries plus one member organization, the European Community
Chair:	Senegal; Headquarters in Rome, Italy
Homepage:	http://www.fao.org

Global Spatial Data Infrastructure (GSDI)

The GSDI is mainly a commercial group promoting a global data infrastructure similar to the regional developments such as the FGDC in the United States. GSDI offers support that includes education in technological, legal, and economic matters.

Scope:	GSDI is an advocacy organization whose goal is to encourage compatible Spatial Data Infrastructures (SDI), globally identifying best practices, and providing a forum for the discussion of capacity building.
Objectives:	
	• Foster active multinational participation in GSDI Working Groups for technology, legal and economic matters.
	• Develop and disseminate an SDI Implementation Guide in all UN approved languages.
	• Educate decision-makers on the benefits of GSDI.
	• Advance the GSDI mission until a global SDI is achieved.
Members:	From more then 50 countries
	Global affiliations:
	• International Steering Committee for Global Mapping (ISCGM)
	• United Nations Geographic Information Working Group (UNGIWG)
	• International Symposium for Digital Earth (DE)
	• Global Disaster Information Network (GDIN)
	• U.S. Federal Geographic Data Committee (FGDC)
	• European Umbrella Organization for Geographic Information (EUROGI)

	• Permanent Committee on Spatial Data Infrastructure for the Americas (PC IDEA)
• Permanent Committee on GIS Infrastructure for the Asia and the Pacific (PCGIAP)
• Open GIS Consortium (OGC)
Chair:	The Netherlands; secretariat in Reston, Virginia, USA
Homepage:	http://www.gsdi.org

IEEE Geoscience and Remote Sensing Society (GRSS)

The GRSS is an international society that brings together experts on remotely sensed data in many fields of application. It is the intention of the society to create a forum within which scientists and engineers involved in a wide range of Earth remote sensing technologies (airborne, ship based, spaceborne, active, passive, optical, microwave, ...) can develop techniques to meet the requirements of a diverse set of data users. The annual International Geoscience and Remote Sensing Symposium (IGARSS) has become one of the most well-attended meetings for the exchange of the ideas and information in Earth remote sensing.

Scope:	The members of GRSS come from both engineering and scientific disciplines. Those with engineering backgrounds are often familiar with geophysics, geology, hydrology, oceanography, and/or meteorology. The fusion of geoscientific and engineering disciplines in projects of global scope give the GRSS a unique interdisciplinary character.
The Society was first known as the Geoscience Electronics Group, formed in 1962.
Chair:	Selected from its International membership, presently USA
Meetings:	Since 1981 it has sponsored the annual International Geoscience and Remote Sensing Symposium (IGARSS) series.
Homepage:	http://www.ewh.ieee.org/soc/grss

International Association of Geodesy (IAG)

International Association of Geodesy

Geodesy supplies the reference systems for geographic information. As a science, geodesy measures the figure of the earth, monitors the earth's rotation and determines the gravity field being required to build elevation reference systems.

Members:	More than 1,000 personal members
Chair:	Italy
Homepage:	http://www.gfy.ku.dk/~iag/

International Cartographic Association (ICA)

The ICA represents the cartographers of the world. The work of the society has presently focused on map design and dedicated maps for special user groups like children and the blind. The ICA has about 100 national, institutional, and company members.

Scope:	ICA is the world authoritative body for cartography, the discipline dealing with the conception, production, dissemination and study of maps
Members:	81 national members
	15 affiliate members (institutions, companies)
Meetings:	International Cartographic Congress, every two years
Chair:	Sweden
Homepage:	http://www.icaci.org

International Civil Aviation Organization (ICAO)

The ICAO is the organization that deals with all international civil aviation questions. The membership includes practically all the countries of the world. The navigation of aircraft is the ICAO topic that generates the most interest in the standardisation of geographic information.

Scope:	The aims and objectives of ICAO are to develop the principles and techniques of international air navigation and to foster the planning and development of international air transport. ICAO was founded in 1944 in Chicago. Its predecessor, the ICAN (International Commission on Air Navigation), was founded in 1910 in Paris.
Members:	188 contracting states
Meetings:	Assembly meets once every three years
Chair:	Lebanon; headquarters in Montreal, Canada
Homepage:	http://www.icao.int

International Federation of Surveyors (FIG)

International Federation of Surveyors
Fédération Internationale des Géomètres
Internationale Vereinigung der Vermessungsingenieure

FIG was founded in 1878 in Paris. It is a federation of national associations and is the only international body that represents all surveying disciplines.

Scope:	Surveyors are professional people whose academic qualifications and post-graduate training enables them to advise on the management and use of both rural and urban land and property whether developed or undeveloped. Surveyors understand the legislation governing land and property, markets, supporting services, and the economics of construction, management, maintenance, acquisition and disposal. FIG aims to ensure that the disciplines of surveying and all who practise them meet the needs of the markets and communities that they serve. It realises its aim by promoting the practice of the profession and encouraging the development of professional standards. FIG is an UN-recognised non-government organization (NGO).
Members:	Representatives from more than 110 countries
Meetings:	Annual congresses
Chair:	Germany
Homepage:	http://www.fig.net

International Hydrographic Bureau (IHB), International Hydrographic Organization (IHO)

The IHO is an organization consisting of hydrographic agencies from most of the maritime countries around the world. The IHO was established in 1921 and supports safety in navigation and protection of the marine environment. One of their major efforts is the creation of an International Standard for digital hydrographical charts.

Scope: The IHO is an intergovernmental consultative and technical organization that was established in 1921 in order to support the safety in navigation and the protection of the marine environment.

Objectives:
- Coordination of the activities of the national hydrographic offices
- Greatest possible uniformity in nautical charts and documents
- Adoption of reliable and efficient methods of carrying out hydrographic surveys
- Development of the sciences in the field of hydrography and oceanography

Members: 70 maritime states
Meetings: International Hydrographic Conference every five years in Monaco
Chair: Greece; International Hydrographic Bureau (IHB) in Monaco
Homepage: http://www.iho.shom.fr

International Society for Photogrammetry and R(

The ISPRS is the traditional international society for photogrammetry and later extended its scope to include remote sensing. The society primarily represents the engineering point of view, gathering all research on sensors (including their models) and on the mathematical models for the derivation of spatial information from imagery. The ISPRS offers expertise on three-dimensional geographic information.

Objectives:	Photogrammetry and remote sensing is the art, science, and technology of obtaining reliable information from non-contact imaging and other sensor systems about the earth and its environment. Established in 1910, ISPRS promotes and coordinates research in the field of photogrammetry and remote sensing.
Members:	More than 100 national societies, more than 50 sustaining members
Meetings:	ISPRS congresses every four years, midterm symposia in between
Chair:	Australia
Homepage:	http://www.isprs.org

International Steering Committee for Global Mapping (ISCGM)

The ISCGM, initiated by Japanese scientists, is a small group that aims at global mapping to improve environmental protection and disaster management.

Objectives:	The ISCGM examines measures that would foster the development of global mapping to facilitate the implementation of global agreements and conventions for: • environmental protection and • mitigation of natural disasters The ISCGM was established in 1996 in Tsukuba, Japan.
Members:	Less than 100 personal members
Chair:	Canada
Homepage:	http://www.iscgm.org/html4/index.html

Joint Research Centre of the European Union (JRC)

The JRC is the research centre of the European Commission. The European Commission functions as the government of the European Union (EU). One of the seven laboratories of the JRC governs the creation of the European geographic data infrastructure (INSPIRE) and is located in Ispra, Italy.

Scope: The JRC is the European Union's scientific and technical research laboratory and is an integral part of the European Commission. The JRC is a Directorate General, providing the scientific advice and technical know-how to support EU policies.
Members: Directorate-General and 7 JRC Institutes
Chair: United Kingdom; Directorate General in Brussels, Belgium
Homepage: http://www.jrc.org

Open GIS Consortium, Incorporated (OGC)

The OGC is the worldwide leading consortium of GIS industries promoting the interoperability of geographic information across platform, system, and country borders. The main field of current activity is the complete integration of the sources of geographic information based on the Internet.

Scope: The OGC is an international industry consortium of more then 230 companies, government agencies and universities participating in a consensus process to develop publicly available geoprocessing specifications. Open interfaces and protocols defined by Open GIS Specifications support interoperable solutions that "geo-enable" the Web, wireless and location based services, mainstream IT, and empower technology developers to make complex spatial information and services accessible and useful with all kinds of applications.
Members: More than 230 companies, government agencies and universities
Chair: USA
Homepage: http://www.opengis.org

Permanent Committee on GIS Infrastructure for Asia and the Pacific (PCGIAP)

The PCGIAP is a regional organization for Asia and the Pacific that coordinates the activities of its member states in the field of geographic information. The major initiatives come from countries like China, Australia, Japan, India, and Iran.

Objectives:	The aims of the Committee are to maximize the economic, social and environmental benefits of geographic information by providing a forum for nations from Asia and the Pacific to:
	• cooperate in the development of a regional geographic information infrastructure,
	• contribute to the development of the global geographic information infrastructure,
	• share experiences and consult on matters of common interest, and participate in any other form of activity such as education, training, and technology transfer.
Members:	55 nations
Chair:	China
Homepage:	http://www.gsi.go.jp/PCGIAP

Permanent Committee on Spatial Data Infrastructure for the Americas (PC IDEA)
(Comité Permanente para la Infraestructura de Datos Geoespaciales de las Américas, CP IDEA)

Initiated by Colombia, the PC IDEA is an organization of the national agencies for mapping, cartography, and geographic information in Latin America including the U.S. and Canada. PC IDEA was established in the year 2000. The Permanent Committee is comprised within the principles of the 21st Agenda of the United Nations Conference on the Environment and Development in order to maximize the economic, social and environmental benefits derived from the use of geospatial information, the knowledge and exchange of experiences and technologies by the different countries, based on a common development model that permits the establishment of a geospatial data infrastructure in the region of the Americas.

Objectives: The PC IDEA helps its member organizations to:
- establish and coordinate the policies and technical rules for the development of a regional data infrastructure for the Americas,
- stimulate the cooperation, investigation, complementation, and the exchange of geographic information,
- define strategies to help the member countries to build their cadastral system.

Members: 24 national agencies for mapping, cartography, and geographic information
Chair: Colombia
Homepage: http://www.cpidea.org.co

Scientific Committee on Antarctic Research (SCAR)

The SCAR coordinates scientific activities regarding the Antarctic. One of their current projects is the creation of a spatial data model for the continent. The SCAR is an organization open to governmental and non-governmental members.

Objectives: The principal objectives of SCAR are:
- to initiate, promote, and coordinate international scientific activity in the Antarctic with a view to framing and renewing scientific programs of circumpolar scope and significance,
- to continue to review scientific matters pertaining to the integrity of the Antarctic environment, including the conservation of its terrestrial and marine ecosystems,
- to provide, upon request, scientific and technological advice to the Antarctic Treaty Consultative Meetings and other organizations, both governmental and non-governmental.

Members: 27 national polar research committees, 11 other members (associated members and international unions)
Chair: Germany

Homepage: http://www.scar.org

UN ECE (United Nations Economical Commission for Europe) Statistical Division

The UNECE is a UN organization promoting a close cooperation of the European statistical bodies according to UN recommendations. Statistical agencies rely on digital geographic information because a lot of statistical data have spatial relations.

Scope: The main objectives of the Conference of European Statisticians are:
- to improve national statistics and their international comparability in regard to the recommendations of the Statistical Commission of the United Nations, the Specialised Agencies and other appropriate bodies as necessary,
- to promote close coordination of the statistical activities in the ECE region of international organizations in order to achieve greater uniformity in concepts and definitions and to minimise the burden on national statistical offices,
- to respond to any emerging need for international statistical co-operation arising out of transition, integration and other processes of cooperation both within the ECE region and between the ECE region and other regions.

Homepage: http://www.unece.org/stats/stats_h.htm

United Nations Geographic Information Working Group (UNGIWG)

The UNGIWG is a U.S. based initiative to promote the use of geographic information for better decision-making.

Scope: The overarching objective of the UNGIWG is to promote the use of geographic information within the United Nations System and Member States for better decision-making

Homepage: http://ungiwg.unep.org

United Nations Group of Experts on Geographical Names (UNGEGN)

The UNGEGN is an UN expert group that provides technical recommendations for the international spelling of geographic names.

Scope: In 1959, ECOSOC (Economic and Social Council) Resolution 715A (XXVII) paved the way for a small group of experts to meet and provide technical recommendations on standardising geographical names at the national and international levels.
Chair: South Africa
Homepage: http://unstats.un.org/unsd/geoinfo/ungegn.htm

World Meteorological Organization (WMO)

The WMO is the worldwide weather organization dealing with weather prediction, climate change and related topics. The membership list of the WMO consists of almost all countries world wide. The interest of the WMO in standardised digital geographic information is obvious.

Scope: From weather prediction to air pollution research, climate change related activities, ozone layer depletion studies and tropical storm forecasting, the WMO coordinates global scientific activity to allow increasingly prompt and accurate weather information and other services for public, private and commercial use, including international airline and shipping industries.
Members: 185 organizations
Homepage: http://www.wmo.ch/indexflash.html
Meetings: World Meteorological Congress every four years
Chair: Nigeria; secretariat in Geneva, Switzerland

5.2.2 Open GIS Consortium (OGC)

5.2.2.1 Background

The Open GIS Consortium (OGC) is currently the world's leading industrial consortium for the implementation of geographic information systems. Mapping of geographic information across the Internet is by far the most important topic of today. In order to enable the contributing systems to communicate smoothly and exchange their information, the OGC is working on a suite of industrial standards. Their spirit and their development process is different from the ISO approach in that the OGC primarily develops Implementation Specifications whereas the ISO standardises the Abstract Specifications on the top level.

Founded in 1994, the OGC evolved from the GRASS user groups. GRASS (Geographic Resources Analysis Support System) is the geographic information system that was developed and maintained by the U.S. military until they decided to switch to commercial products of ESRI.

Today, the OGC has well over 200 members representing the complete spectrum of players on the GIS market place. Their background ranges from system manufacturers across production companies and government agencies to universities and research laboratories. The membership type is split into four main categories: Strategic, Principal, Technical, and Associate. The annual fee is different for each category: Strategic Members $250,000 US, Principal Members $50,000 US, Technical Members $10,000 US, and Associate Members $4,000 US, or $300 US in the case of universities. In practice, the annual fee of Strategic Members is negotiable; because the fee is often paid by installing a work position on the staff level and/or by sponsoring larger implementation projects.

The following table lists the eight Strategic Members. Their contribution covers more then 50% of the annual budget of the OGC.

Table 5.1. List of Strategic Members of the Open GIS Consortium

Company	Field of Activity	
BAE Systems	Information and Electronic Warfare Systems	U.S.
General Dynamics	Mission-critical information systems, aviation, shipbuilding	U.S.
Intergraph	Mapping and Geospatial Solutions	U.S.
Lockheed Martin	Space systems	U.S.
Northrop Grumman	Electronics, IT, mission systems, ships, space systems	U.S.
FGDC	Federal Geographic Data Committee	U.S.
NASA	National Aeronautics and Space Administration	U.S.
NIMA	National Imagery and Mapping Division	U.S.

Though the Strategic Members essentially pay the bill, a great number of the other members frequently contribute innovative technologies.

The OGC has established a formal framework for their development of the standards that consists of a hierarchy of committees, a defined voting procedure and a number of formalized document types. OGC has adopted the ISO template for the purpose of publishing a printed standard.

The document types are called Request for Proposal (RFP), Request for Information (RFI), Request for Comment (RFC), and Request for Technology (RFT).

An RFP is issued to the public by the OGC after a piece of the Open GIS Specification has been agreed upon. RFP submissions offered by members are demonstrably workable Implementation Specifications. An RFI is issued to the public by the OGC at critical points to ensure that all industry players have an opportunity to give the OGC their wisdom during the development of a specification. An RFC describes interface technology offered by a member or team of members who wants the consortium to adopt their technology rather than develop similar technology through the usual Technical Committee process. An RFT is issued to the public at the beginning of an Interoperability Initiative to stimulate submissions of interface technologies that need to be evaluated and modified in a testbed (OGC 2003).

The committees involved in the OGC's standardisation process are structured in three levels. The strategic decisions are made in the Management Committee (MC). The technical work is coordinated in the Technical Committee (TC). The Working

Clifford A. Kottman Ph.D. studied Mathematics with focus on Physics and Philosophy at the Loyola University in Los Angeles and with focus on Analysis and Topology at the University Iowa, where he also received his Ph.D. degree. One of his widest known developments is the Digital Chart of the World. Dr. Kottman conceived, obtained funding for, and planned this project.

Before he joined the OGC he had various leading positions in companies like Intergraph and institutions like the National Imagery and Mapping Agency (NIMA) or its predecessors. In the OGC Dr. Kottman has led the Technical and Planning Committees toward the establishment of global consensus Implementation Specifications in the arena of geospatial information and services.

The academic world was never far from Dr. Kottman's domain. He has served in various functions including tenured Associate Professor of Mathematics at a number of academic institutions. The locations of the universities range from Oregon and Iowa to Louisiana and the Washington D.C. area.

Dr. Kottman's first interaction with standardisation came in 1978 with the introduction of digital geographic products in support of the U.S. military. Among his many positions he has been the liaison from the Open GIS Consortium to ISO/TC211. He was instrumental in the drafting and negotiations that led to the Cooperative Agreement between OGC and ISO/TC211.

Groups (WGs) and the Special Interest Groups (SIGs) deal with specific fields of work and individual specifications.

The business plan developed by the Management Committee must be approved by the OGC Board of Directors (BOD). This group is comprised of respected leaders in the Information Technology community but they need not be affiliated with OGC member organizations. The Strategic Members make recommendations to the Board of Directors concerning strategic opportunities for the consortium.

The Working Group members draft the Requests for Information and/or Proposals (RFIs and RFPs) and evaluate the proposed technologies in response to the RFPs. The Working Group members also evaluate technologies proposed by OGC Members as Requests for Comments (RFCs) or they work on the refinement of the Abstract Specification.

Voting in WGs is by simple majority of OGC Members present at the WG meeting, with the caveat that no OGC member organization may cast more than one vote.

The duties of the Special Interest Groups (SIGs) are comparable with those of Working Groups, with a stronger focus on Abstract Specifications and editing work for the Technical Committee (TC). The SIG has the same voting structure as that of a Working Group.

Presently the OGC has the following Working Groups (WGs) and Special Interest Groups (SIGs). The chairpersons are mostly representatives of the Strategic and Principle OGC Members.

Table 5.2. Working Groups (WGs) and Special Interest Groups (SIGs) of the OGC

Title	Chair
Architecture SIG	Federal Geographic Data Committee
Catalog Revision WG	Federal Geographic Data Committee
Coordinate Transformation WG	Shell
Coordinate Transformation Revision WG	CADCorp.
Coverages WG	Oracle
Grid Coverages Revision WG	PCI Geomatics
Decision Support SIG	U.S. Army Corps of Engineers
Defense and Intelligence SIG	OGC
Disaster Management SIG	USGS, Oracle
Earth Observation SIG	PCI Geomatics, International Interfaces
Feature SIG	Oracle
Geography Markup Language (GML) SIG	Galdos Systems, NASA/CSC
Image Exploitation Services SIG	BAE Systems
Location Services SIG	Oracle
Metadata SIG	ESRI
Natural Resources and Environmental SIG	Ecosystem Associates
Open Spatial Publishing and Discovery Infrastructure (OSPDI) WG	SRI
Information Communities and Semantics SIG	Image Matters, Lockheed Martin
Simple Features for CORBA Revision WG	SICAD Geomatics
Telecommunications SIG	OGC
WWW Mapping SIG	International Interfaces

The Technical Committee (TC) develops and maintains the Open GIS Specifications and is focused on both general and domain specific interface development. The Technical Committee is composed of the Technical Committee Chair (appointed by the Board of Directors), representatives of all member organizations of the OGC and other individuals deemed appropriate by the Board of Directors or the Management Committee. The right to vote within the Technical Committee is only granted to the representatives of Strategic, Principal, and Technical Members. Only one Voting TC Member may vote on behalf of each such Member organization at TC meetings.

The OGC Management Committee (MC) is composed mainly of Strategic and Principal Member representatives and provides guidelines and a management structure for OGC's Technical Committee and Interoperability Program. The OGC Management Committee approves special negotiated memberships and committee participation. It has asked ISO/TC211, for example, to be a non-paying voting member of the Management Committee.

5.2.2.2 Program of work

The vision of the OGC is a world in which everyone benefits from geographic information and services that are made available across any network, application, or platform. Through the tremendous efforts of the OGC in the past, some of the visionary developments, such as web mapping, have made considerable progress. The OGC has coined the term "interoperability" to describe the technical pre-requisites for the realisation of their vision.

The program of work is structured in two categories, the Specification Program and the Interoperability Program.

Guided by the OGC Technical Committee, the work in the Specification Program culminates in the publication of Specifications, sometimes called OGC standards. The Specifications again fall into two categories, the Abstract Specifications and the Implementation Specifications. Originally, the concept of the OGC Specification Program was aimed at building a complete suite of GIS standards. Today, this ambitious goal has been unofficially modified towards a focus on the Implementations Specifications. The work on the Abstract Specfication has become the concern of the ISO/TC211 (Geographic information / Geomatics). Some of the original OGC Abstract Specifications have become the working drafts for the ISO-standardisation.

The Interoperability Program is a series of engineering initiatives aimed at accelerating the development and acceptance of OGC Specifications. The Interoperability Program has evolved into the OGC's main field of activity.

The program of work of the OGC is not as clearly structured as the ISO 19100 family of GIS-standards. This is mainly a result of the project-driven approach of the Interoperability Program and of some Specification developments that often depend on limited sponsorship. Therefore, while it is currently focused on web mapping, the program of work is continuously undergoing slight changes. Two of the better-known Implementation Specifications were transferred to the ISO standardisation process and are known as the "Web Map Service Interfaces Implementation Specification" and the "Geography Markup Language (GML) Implementation Specification".

Examples for initiatives of the Interoperability Program include "Web Mapping Testbeds", the "Geospatial Fusion Services Pilot Project", and the German "North Rhine Westphalia Pilot Project (NRWPP)". The role of OGC in these projects was to coordinate the development work of their members.

5.2.2.3 Relation of the OGC to ISO/TC211

The OGC and the ISO/TC211 (Geographic information / Geomatics) were established almost simultaneously in the early nineties. Both were aimed at the international standardisation of geographic information and both accomplished it with different backgrounds in tradition and thinking. The OGC is a mainly US-based industry-oriented consortium. The ISO/TC211 was proposed by Canada in order to link the European traditions with work in other parts of the world. The Open GIS Consortium aims at the efficient development of joint solutions among the members of the consortium while the goal of ISO-standards is long term, generic solutions developed under a democratic participation of all interested parties.

The concurrent development of two independent suites of standards in the same domain is not feasible for practical solutions. In addition, it means a heavy waste of resources because one of the two standards will sooner or later fall out favour. Overall, the field of geographic information is small compared to other businesses such as the automotive industry or the money market.

Originally, the OGC was only an external liaison member to the ISO/TC211 as were many others. The OGC and the ISO/TC211 formally signed a cooperative agreement in 1998 wherein the ISO/TC211 adopted OGC specifications and published them as ISO-standards. Conversely, the OGC has the right to publish the ISO-standards under the OGC-cover. In order to monitor the matters rising from the agreement, a co-ordination group was established. This group facilitates the smooth running of the agreement, monitors the liaison on the work item level, and functions as a clearinghouse.

In practice the alignment of the work of both organizations is not a simple task. The intentions of both organizations and the formal process of the standard's development are not fully compatible. A simple, but not always applied, border separates the abstract standards (being the domain of ISO) from the Implementation Specifications (being the domain of the OGC).

The ISO-19100 standards that are not abstract have been adopted from the OGC. They are the ISO 19128 (Web Map server interface), the ISO 19136 (Geographic Markup Language), and the ISO 19139 (Metadata – Implementation Specification). It is sometimes questioned whether ISO should use its name for non-abstract standards. Most of the Implementation Specifications are living documents that evolve according to the users' needs. It might happen that the International Standard being published by ISO becomes outdated by the time the formal processes are complete. During that time, the OGC and their member companies may already have considerably advanced the implementation and the documentation in a later version.

The cooperative agreement aligns the milestones of the standard developments of ISO and of the OGC. The following table depicts only an approximate relationship because the concepts of the development stages do not precisely match.

Table 5.3. Relation between ISO and OGC milestones

organization	milestones				
ISO	NWIP	CD	DIS	FDIS	IS
OGC	RFP	V 0.0	V 1.0	V 2.0	

NWIP = New Work Item Proposal
CD = Committee Draft
DIS = Draft International Standard
FDIS = Final Draft International Standard
IS = International Standard
RFP = Request for Proposal

Previous experiences have shown that the most efficient way of transferring an industry standard to an ISO-standard is by waiting until the industry has finalized their consensus-building process. A similar procedure is deemed advisable for OGC Specifications intended to become ISO-standards.

6 Applications

6.1 Canadian GIS industry

Canada has a strong geomatics industry and many Canadian companies have contributed innovative solutions to the world market of GIS technology. Galdos Systems, CubeWerx, Compusult, and PCI Geomatics are presently the best-known representatives. Many of their products were developed in close cooperation with companies and institutions in the USA.

Galdos Systems Inc. is well known as the home of the Geography Markup Language (GML). The father of GML, Ron Lake, is the president and CEO of Galdos Systems. The company was founded in 1992 and is located in Vancouver, British Columbia. Galdos Systems contributed to the development of the ISO 19136 (Geography Markup Language) and the ISO 19118 (Encoding). The Open GIS Consortium has also adopted the GML as the global encoding standard for geospatial data.

CubeWerx Inc. has contributed a number of innovative developments in the Web Mapping field. The company has invented the Cascading Web Map server that allows a chaining of web mapping services. CubeWerx was founded in 1996 and is located in Gatineau, Québec, near Ottawa. It contributed to the ISO 19128 (Web Map server interface) and the ISO 19119 (Services). CubeWerx is also involved in many OGC-developments.

Compusult Ltd. is a specialist in metadata. The company has developed the Meta Manager that is used by the U.S. Federal Geographic Data Committee. Compusult was founded in 1985 and is located in Mount Pearl, Newfoundland. Their involvement in the ISO/TC211 works covers the ISO 19115 (Metadata) and the ISO 19110 (Methodology for feature cataloguing).

PCI Geomatics has developed a geographic information system with a rich functionality, especially in the field of remote sensing. The company is located in Toronto, Ontario. PCI Geomatics has participated in the development of the Web Coverage Service. They were involved in the development of the ISO 19123 (Schema for coverage geometry and functions).

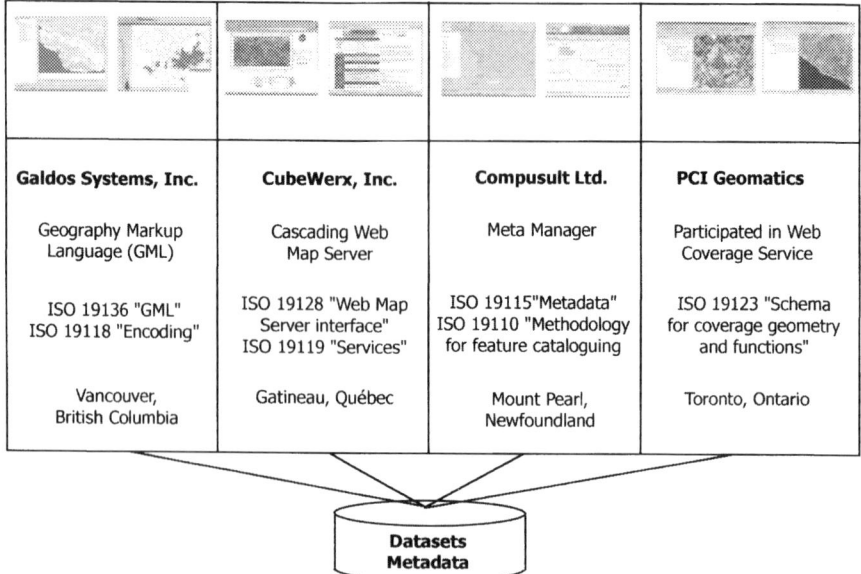

Fig. 6.1. Canadian companies contributing to the work of ISO/TC211

6.2 German standard-based systems for cadastral and topographic information

With the emergence of international standardisation for digital geospatial information first in CEN/TC287 and later in ISO/TC211 the Working Committee of the Surveying Authorities of the States of the Federal Republic of Germany (AdV) decided that legacy cadastral and topographic information systems (e.g. ALK, ALB, ATKIS) had to undergo a complete redesign to embrace International Standards. Members of AdV took an active part in Germany's national body DIN alongside with German GIS vendors shaping standards for digital geospatial information in the respective CEN and ISO Technical Committees and at the same time applying them to the design of next generation highly integrated German cadastral and topographic information systems AFIS, ALKIS, ATKIS. Of these ALKIS (Authoritative Cadastral Map Information System) is internationally best known as an approach to challenge practicality of ISO 19100 series standards that can serve as a testbed and reference implementation for high quality cadastral data management aimed at positional and timely accuracy and data and referential integrity.

The modelling of AFIS-ALKIS-ATKIS was done in UML applying ISO/TS 19103 (Conceptual schema language) and ISO 19109 (Rules for application schema) to the cadastral and topographic application domain. The UML model is tightly integrated with ISO/TC211 Harmonized UML Model using Rational Rose as a tool to

6.2 German standard-based systems for cadastral and topographic information

ensure consistency across the UML model. Using this approach, which can be charaterized as Model Driven Architecture (MDA), application schemata could make use of all predefined data types, types and features existing in the Harmonized UML Model deploying them as base classes for generalization / specialisation relationships as well as for typing roles in associations and aggregations or attributes and operation parameters in respective cadastral application feature classes.

Since AFIS-ALKIS-ATKIS is envisioned by German AdV as an integration platform for other geospatial applications related to the cadastral and topographic application domain a AFIS-ALKIS-ATKIS Base Schema was factored out to be potentially reused in application schemata other then AFIS-ALKIS-ATKIS.

Representation of geometric and topological properties of feature classes in AFIS-ALKIS-ATKIS is completely based on ISO 19111 (Spatial referencing by coordinates) and ISO 19107 (Spatial schema) using also its informative Simple Topology examples which show the integration of geometric and topological properties within one application feature class, allowing for easier implementation based on existing GIS base products. AFIS-ALKIS-ATKIS is actually using of profile of ISO 19107 (Spatial schema) restricting the variety of possible geometric and topological representations for features and introducing a new kind of topological theme where only border lines of features with surface geometry take part in a topological complex.

Metadata used to describe quality aspects of single features and feature sets, respectively as well as metadata for organizational aspects of geodata capturing, management, and distribution are modelled based on ISO 19115 (Metadata) and its extension methodology.

Catalogues defining the semantics, constraints and structural and behavioural aspects of feature types are generated for AFIS, ALKIS, and ATKIS application schemata according to ISO 19110 (Methodology for feature cataloguing). AFIS-ALKIS-ATKIS base schema extends the 19110 part of ISO/TC211 Harmonized UML Model and XML encoded feature catalogues are generated from the AFIS-ALKIS-ATKIS UML model and its associated documentation.

To bridge the gap between AFIS-ALKIS-ATKIS base and application schemata deploying ISO 19100 series standards captured in ISO/TC211 Harmonized UML Model, which abstract from implementation aspects, and requirements for specifications aimed at interoperability of AFIS-ALKIS-ATKIS implementations based on GIS products the use of ISO 19118 (Encoding) is of critical importance. A set of encoding rules have been defined to derive XML Schema Definitions for AFIS-ALKIS-ATKIS base and application schemata as well as ISO 19100 standards from respective UML models, following a Model Driven Architecture (MDA) approach.

Since interoperability of AFIS-ALKIS-ATKIS COTS implementations is essential to build up viable geodata infrastructures e.g. in the German state of North Rhine Westphalia (GDI NRW) AdV decided to incorporate Open GIS Consortium (OGC) Implementation Specifications i.e. Geography Markup Language 3.0 (GML), Web Feature Server (WFS), and Filter Encoding into its Standard-based data exchange interface (NAS). NAS consists of a set of well defined operations encoded as XML Schema Definitions for matching request and response pairs. XML Schema Definitions for WFS, Filter Encoding, GML 3.0, and AFIS-ALKIS-ATKIS base and application schema based on GML 3.0 are incorporated in NAS.

NAS Operations provide means for standardised client server communication in business processes of cadastral and surveying authorities and their information exchange with e.g. consultant surveying engineers and customers of cadastral and topographic map and/or feature data. Products comprise XML encoded feature datasets i.e. feature datasets to perform regular updates of feature data stored at customer sites, map datasets including cartographic presentation data together with feature data, authoritative reports i.e. change reports of legal relevance to land ownership, and other authoritative documents required by customers. Specific NAS Operations allow consultant surveying engineers, authorized to perform cadastral surveying tasks, to send feature data updates to AFIS-ALKIS-ATKIS data server located at cadastral and surveying agencies.

ISO/TC211 and OGC have identified the need to harmonize ISO 19118 (Encoding) and GML 3.0 and due to significance of harmonisation for long-term success of XML based interoperability of GIS applications like German cadastral and topographic information systems AFIS-ALKIS-ATKIS German DIN members play an important role in resolving issues related to ISO 19136 (Geography Markup Language).

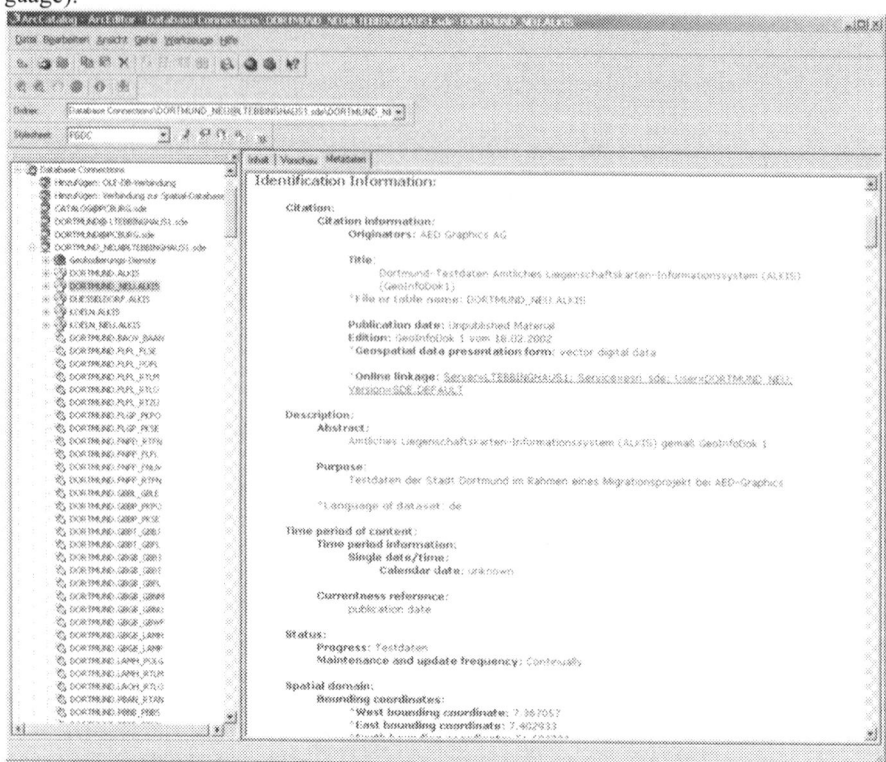

Fig. 6.2. ALKIS-metadata in ArcCatalog based on ISO 19115 (Metadata)

The chapter 6, section 2 was provided by Jürgen Ebbinghaus (Ebbinghaus 2003).

Annexes

Bibliography

Abadi M, Cardelli L (1996) A Theory of Objects. Springer, New York
Ahuja RK, Magnanti TL, Orlin JB (1993) Network Flows: Theory, Algorithms and Applications. Prentice Hall, Englewood Cliffs, New Jersey
Albertz J, Kreiling W (1989) Photogrammetric Guide. Wichmann, Karlsruhe, Germany
Arnold DB, Duce DA (1990) ISO Standards for Computer Graphics, The First Generation. Butterworths
Bartels RH, Beatty JC, Barsky BA (1987) An Introduction to Splines for Use in Computer Graphics & Geometric Modelling. Morgan Kaufmann, San Francisco
Bellman RE (1958) On a routing problem. Quarterly of Applied Mathematics, vol 16: 87-90, Brown University Press
Booch G, Jacobson I, Rumbaugh J (1999) UML User Guide. Addison-Wesley, Boston
Bray T, Paoli J, Sperberg-McQueen CM (eds) (1998) Extensible Markup Language 1.0. W3C Recommendation, 10 February 1998, <http://www.w3.org/TR/1998/REC-xml-19980210>
Brozio S (2003) Mathematical surfaces. Personal communication, designed with Tablecurve, FH Neubrandenburg, Germany
BTS98 (2001) Delay-Volume Relations for Travel Forecasting: Based on the 1985 Highway Capacity Manual. U.S. Departement of Transportation, Washington D.C., <http://tmip.fhwa.dot.gov/clearinghouse/docs/general/dvrt/intro.stm>
Cherkassky BV, Goldberg AV, Radzik T (1993) Shortest Paths Algorithms: Theory and Experimental Evaluation. Technical Report 93-1480, Computer Science Department, Stanford University
Christensen E, Curbera F, Meredith G, Weerawarama S (2001) Web services Description Language (WSDL) 1.1, W3C Note, <http://www.w3.org/TR/wsdl>
Clapham C (1996) Concise Dictionary of Mathematics. 2nd edn, Oxford University Press, Oxford
Clementini E, Di Felice P (1995) A Comparison of Methods for Representing Topological Relationships. Information Sciences 3, Elsevier, Amsterdam
Clementini E, Di Felice P (1996) A Model for Representing Topological Relationships Between Complex Geometric Features in Spatial Databases. Information Sciences 90 (1-4):121-136, Elsevier, Amsterdam
Clementini, E, Di Felice P, van Oostrom P (1993) A Small Set of Formal Topological Relationships Suitable for End-User Interaction. In: Abel D, Ooi BC (eds) Advances in Spatial Databases — Third International Symposium, SSD '93, LNCS 692: 277-295, Springer, Singapore

CNIG (1995) Nomenclature d'échange du CNIG associée à la norme EDIGéO. Conseil National de l'Information Géographique (CNIG), <http://www.cnig.fr>

Cook S, Daniels J (1994) Designing Object Systems - Object-Oriented Modelling with Syntropy. Prentice Hall, Englewood Cliffs, New Jersey

Cox S, Daisey P, Lake R, Portele C, Whiteside A (2003) OpenGIS Geography Markup Language (GML) Implementation Specification, version 3.00. Open GIS Consortium, Document 02-023r4, Wayland, Massachusetts

Curlander JC, McDonough RN (1991) Synthetic Aperture Radar, Systems and Signal Processing. Wiley, Hoboken, New Jersey

Daintith J, Nelson RD (1989) Dictionary of Mathematics. Penguin Books, London

Dale P (2000) ISO/TC211 or "A Different Route". Geo-Informatics 7(3), Emmeloord, Netherlands

Danko D (2002) Implementing ISO Metadata. Presentation at the ISO/TC211 workshop standards in action in Gyeongju, South Korea

De La Beaujardière J (ed.) (2002) Web Map Service Implementation Specification 1.1.1. OGC Document 01-068r3, Wayland, Massachusetts

Dershowitz N, Reingold EM (1997) Calendrical calculations. Cambridge University Press, Cambridge

DIGEST (2003) Digital Geographic Information Exchange Standard, <http://www.digest.org>

Dijkstra E (1959) A note on two problems in connection with graphs. Numerische Mathematik, vol 1: 269-271, Springer

DIN95 (1995) DIN 18716-1, Photogrammetry and remote sensing - Part 1: General terms and specific terms of photogrammetric data acquisition. Beuth, Berlin

DIN96a (1996) DIN18716-2, Photogrammetry and remote sensing - Part 2: Specific terms of photogrammetric data acquisition. Beuth, Berlin

DIN96b (1996) DIN18716-3, Photogrammetry and remote sensing - Part 3: Remote sensing terms. Beuth, Berlin

DOM (1998) Document Object Model Level 1 Specification, Version 1.0. W3C Recommendation 01-October-1998, <http://www.w3.org/TR/REC-DOM-Level-1>

Ebbinghaus J (2003) German standard-based systems for cadastral and topographic information. Personal communication, SICAD GEOMATICS, München

Egenhofer MJ, Herring J (1990) A mathematical framework for the definition of topological relationships. Proceedings of the Fourth International Symposium on Spatial Data Handling: pp 803-813, Zürich

Egenhofer MJ, Herring J (1991) Categorizing binary topological relationships between regions, lines and points in geographic databases. Technical Report, Department of Surveying Engineering, University of Maine, Orono

Elachi C (1988) Spaceborne Radar Remote Sensing: Applications and Techniques. IEEE Press, New York

EPSG (2003) Geodesy Parameters, Version 6.3. European Petroleum Survey Group, <http://www.epsg.org>

FGDC (1998) Content standard for digital geospatial metadata (revised June 1998). Federal Geographic Data Committee, Washington D. C., <http://www.fgdc.gov>

Foley JD, van Dam A, Feiner SK, Hughes JF (1987) Computer Graphics, Principles and Practice. Addison-Wesley, Boston

Ford LR Jr. (1956) Network Flow Theory. Paper P-923, RAND Corporation, Santa Monica, California

Fowler, M (1997) UML Distilled. Addison-Wesley, Boston

Fricker P (2001) ADS40 – Progress in digital aerial data collection. Photogrammetric Week '01, Wichmann, Heidelberg

GEBCO (1991) General Bathymetric Chart of the Oceans. IHO Publication B-7, GEBCO Guidelines, International Hydrographic Bureau, Monaco

Ghiladi V (2002) Do standards users get what they want from ISO standards? ISO Bulletin vol 33 (11): 2

Gurtner W (1998) RINEX: The Receiver Independent Exchange Format, Version 2. Astronomical Institute, University of Berne, September 1998 <ftp://igscb.jpl.nasa.gov/igscb/data/format/rinex2.txt>

Hanisch F (1996/97) Applet for B-Spline curves, Tübingen, <http://www.cs.technion.ac.il/~cs234325/Homepage/Applets/applets/bspline/GermanApplet.html>

Hartlieb B (ed) (2000) Gesamtwirtschaftlicher Nutzen der Normung. published by DIN, Beuth, Berlin

Heller D, Fox S (1994) Proposed ECS Core Metadata Standard. Release 2.0, Hughes Applied Information Systems, 420-TP-001-005, Landover, Maryland

Hinz A, Dörstel C, Heier H (2001) DMC – The Digital Sensor Technology of Z/I-Imaging. Photogrammetric Week '01, Wichmann, Heidelberg

Hudon Y (ed) (2002) Compliance of ISO 19115 and the draft of Canadian core metadata recommended profile with ANSI/NISO Z39.85-2001. Canadian expert group on metadata, Standards Council of Canada

Hudon Y (2003) Cultural and linguistic adaptability. Personal communication, Gouvernement du Québec, Ville de Québec, Canada

Iakovleva, R (1996) Suggestions on the Methodology for Developing the Catalogue of Geographical Objects. Federal Service of Geodesy and Cartography, Moscow

Illingworth V (1997) Dictionary of Computing, Fourth edition, Oxford University Press, Oxford

IEC (2003) http://www.iec.ch

ISO01 (2003) http://www.iso.ch

ISO02 (2003) ISO – Standards Developer's Information Site, <http://www.iso.ch/iso/en/ISOOnline.frontpage>, For standards developers

ISO03 (2003) http://www.tc176.org

ISO04 (2003) http://www.isotc211.org

ISO05 (1992) ISO 31-2:1992, Quantities and units — Part 2: Periodic and related phenomena

ISO06 (1998) ISO 639-2:1998, Codes for the representation of names of languages — Part 2: Alpha-3 code

ISO07 (1996) ISO 690:1996, Documentation — Bibliographic references — Content, form and structure

ISO08 (2000) ISO 704:2000, Terminology work — Principles and methods

ISO09 (1996) ISO 860:1996, Terminology work — Harmonization of concepts and terms

ISO10 (2000) ISO 1087-1:2000, Terminology work — Vocabulary — Part 1: Theory and application

ISO11 (2002) ISO 2859-4:2002, Sampling procedures for inspection by attributes – Part 4: Procedure for assessment of declared quality levels, parts 0 – 3 published between 1985 and 1999

ISO12 (1997) ISO 3166-1:1997, Codes for the representation of names of countries and their subdivisions — Part 2: Country code

ISO13 (1998) ISO 3166-2:1998, Codes for the representation of names of countries and their subdivisions — Part 2: Country subdivision code

ISO14 (1993) ISO 3534-2:1993, Statistics — Vocabulary and symbols — Part 2: Statistical quality control

ISO15 (1989) ISO 3951:1989, Sampling procedures and charts for inspection by variables for percent nonconforming

ISO16 (1983) ISO 6709:1983, Standard representation of latitude, longitude, and altitude for geographic point locations, <http://www.iso.ch>
ISO17 (2000) ISO 8601:2000, Data elements and interchange formats — Information interchange — Representation of dates and times
ISO18 (1988) ISO 8805:1988, Information processing systems – Computer graphics – Graphical Kernel System for Three Dimensions (GKS-3D) functional description
ISO19 (1986) ISO 8879:1986, Information processing — Text and office systems — Standard Generalized Markup Language (SGML)
ISO20 (2000) ISO 9000 (all parts):—, Quality management and quality assurance standard
ISO21 (1987) ISO/TR 9007:1987, Information processing systems — Concepts and terminology for the conceptual schema and the information base
ISO22 (1995) ISO 10005:1995, Quality management — Guidelines for quality plans
ISO23 (1997) ISO 10006:1997, Quality management — Guidelines to quality in project management
ISO24 (1995) ISO 10007:1995, Quality management — Guidelines for configuration management
ISO25 (1992) ISO 10012-1:1992, Quality assurance requirements for measuring equipment — Part 1: Metrological confirmation system for measuring equipment
ISO26 (1997) ISO 10012-2:1997, Quality assurance for measuring equipment — Part 2: Guidelines for control of measurement of processes
ISO27 (1995) ISO 10013:1995, Guidelines for developing quality manuals
ISO28 (1998) ISO/TR 10014:1998, Guidelines for managing the economics of quality
ISO29 (1999) ISO 10015:1999, Quality management - Guidelines for training
ISO30 (1992) ISO 10241:1992, International terminology standards — Preparation and layout
ISO31 (1993) ISO 11180:1993, Postal addressing
ISO32 (1998) ISO/FDIS 12639:1998, Graphic technology — Prepress digital data exchange — Tag image file format for image technology (TIFF/IT)
ISO33 (1999) ISO/TS 16949:1999, Quality systems — Automotive suppliers — Particular requirements for the application of ISO 9001:1994
ISO34 (2002) ISO 19101:2002, Geographic information — Reference model
ISO35 (2001) ISO/PDTS 19103:2001, Geographic information — Conceptual schema language
ISO36 (2002) ISO/DIS 19104:2002, Geographic information — Terminology
ISO37 (2000) ISO 19105:2000, Geographic information — Conformance and testing
ISO38 (2002) ISO/DIS 19106:2002, Geographic information — Profiles 1)
ISO39 (2003) ISO 19107:2003, Geographic information — Spatial schema
ISO40 (2002) ISO 19108:2002, Geographic information — Temporal schema
ISO41 (2002) ISO/DIS 19109:2002, Geographic information — Rules for application schema
ISO42 (2001) ISO/DIS 19110:2001, Geographic information — Methodology for feature cataloguing
ISO43 (2003) ISO 19111:2003, Geographic information — Spatial referencing by coordinates
ISO44 (2001) ISO/DIS 19112:2001, Geographic information — Spatial referencing by geographic identifiers 1)
ISO45 (2002) ISO 19113:2002, Geographic information — Quality principles
ISO46 (2001) ISO/DIS 19114:2001, Geographic information — Quality evaluation procedures 1)
ISO47 (2003) ISO 19115:2003, Geographic information — Metadata
ISO48 (2002) ISO/DIS 19116:2002, Geographic information — Positioning services 1)
ISO49 (2002) ISO/DIS 19117:2002, Geographic information — Portrayal 1)
ISO50 (2002) ISO/DIS 19118:2002, Geographic information — Encoding 1)
ISO51 (2002) ISO/DIS 19119:2002, Geographic information — Services

ISO52 (2001) ISO/TR 19120:2001, Geographic information — Functional standards
ISO53 (2002) ISO/CD 19123:2002, Geographic information — Schema for coverage geometry and functions 1)
ISO54 (2000) ISO/DIS 19125-1:2000, Geographic information — Simple feature access – Part 1: Common architecture 1)
ISO55 (2000) ISO/DIS 19125-2:2000, Geographic information — Simple feature access – Part 2: SQL option 1)
ISO56 (2002) ISO/WD 19129:2002, Geographic information — Imagery, gridded and coverage data framework 1)
ISO57 (2001) ISO/WD 19130:2003, Geographic information — Sensor and data models for imagery and gridded data 1)
ISO58 (2003) ISO/CD 19133:2003, Geographic information — Location based services tracking and navigation 1)
ISO59 (1998) ISO 23950:1998, Information and documentation — Information retrieval (Z39.50) — Application service definition and protocol specification
ISOIEC01 (1994) ISO/IEC Directives:1994, Procedures for the Technical Work of ISO/IEC JTC 1 on Information Technology (third edition)
ISOIEC02 (1997) ISO/IEC Directives, Part 1:1997, Procedures for the technical work, annex N - Registration Authorities.
ISOIEC03 (1997) ISO/IEC Directives, Part 2:1997, Methodology for the development of International Standards, annex E - designation of internationally standardized items
ISOIEC04 (1996) ISO/IEC Guide 2:1996, Standardisation and related activities – general vocabulary.
ISOIEC05 (1998) Report of the JTC1 Ad Hoc meeting of the Technical Direction on Cultural and Linguistic Adaptability and User Interfaces. ISO/IEC JTC1 document N5629
ISOIEC06 (1994) ISO/IEC 2022:1994, Information technology — Character code structure and extension techniques
ISOIEC07 (1993) ISO/IEC 2382-1:1993, Information technology — Vocabulary — Part 1: Fundamental terms
ISOIEC08 (1999) ISO/IEC 2382-4:1999, Information technology — Vocabulary — Part 4: Organization of data
ISOIEC09 (1994) ISO/IEC 7942-1:1994, Information technology – Computer graphics and image processing – Graphical Kernel System (GKS) – Part 1: Functional description
ISOIEC10 (1994) ISO/IEC 8211:1994, Information technology — Specification for a data descriptive file for information interchange
ISOIEC11 (1992) ISO/IEC 8632:1992, Information technology — Computer graphics — Metafile for the storage and transfer of picture description information
ISOIEC12 (1995) ISO/IEC 8824:1995, Information technology — Abstract Syntax Notation One (ASN.1)
ISOIEC13 (2001) ISO/IEC 8859 (all parts):—, Information technology — 8-bit single-byte coded graphic character sets
ISOIEC14 (1992) ISO/IEC 9075:1992, Information technology — Database languages — SQL
ISOIEC15 (1997) ISO/IEC 9592-1:1997, Information technology – Computer graphics and image processing – Programmer's Hierarchical Interactive Graphic System (PHIGS) – Part 1: Functional description
ISOIEC16 (1991) ISO/IEC 9636-1: 1991, Information technology – Computer graphics – Interfacing techniques for dialogues with graphical devices (CGI) – Functional specification – Part 1: Overview, profiles, and conformance
ISOIEC17 (1994) ISO/IEC 9646-1:1994, Information technology — Open Systems Interconnection — Conformance testing methodology and framework — Part 1: General concepts

ISOIEC18 (1994) ISO/IEC 9646-2:1994, Information technology — Open Systems Interconnection — Conformance testing methodology and framework — Part 2: Abstract Test Suite specification
ISOIEC19 (1994) ISO/IEC 9646-5:1994, Information technology — Open Systems Interconnection — Conformance testing methodology and framework — Part 5: Requirements on test laboratories and clients for the conformance assessment process
ISOIEC20 (1994) ISO/IEC 9973:1994, Information technology — Computer graphics and image processing — Procedures for registration of graphical items
ISOIEC21 (1998) ISO/IEC TR 10000-1:1998, Information technology — Framework and taxonomy of International Standardized Profiles — Part 1: General principles and documentation framework
ISOIEC22 (1998) ISO/IEC TR 10000-2:1998, Information technology — Framework and taxonomy of International Standardized Profiles — Part 2: Principles and taxonomy for OSI profiles
ISOIEC23 (1998) ISO/IEC TR 10000-3:1998, Information technology — Framework and taxonomy of International Standardized Profiles — Part 3: Principles and taxonomy for Open System Environment Profiles
ISOIEC24 (1990) ISO/IEC 10027:1990, Information technology — Information Resource Dictionary System (IRDS) framework
ISOIEC25 (1994) ISO/IEC 10303-11:1994, Industrial automation systems and integration — Product data representation and exchange — Part 11: Description methods: The EXPRESS language reference manual
ISOIEC26 (1993) ISO/IEC 10641:1993, Information technology — Computer graphics and image processing — Conformance testing of implementations of graphics standards
ISOIEC27 (2000) ISO/IEC 10646-1:2000, Information technology — Universal Multiple-Octet Coded Character Set (UCS) — Part 1: Architecture and Basic Multilingual Plane
ISOIEC28 (2001) ISO/IEC 10646-2:2001, Information technology — Universal Multiple-Octet Coded Character Set (UCS) — Part 2: Supplementary Planes
ISOIEC29 (1998) ISO/IEC 10746-1:1998, Information technology — Open Distributed Processing – Reference Model: Overview
ISOIEC30 (1994) ISO/IEC 10918:1994, Information technology — Digital compression and coding of continuous-tone still images
ISOIEC31 (1998) ISO/IEC TR 11017:1998, Information technology — Framework for internationalization
ISOIEC32 (2000) ISO/IEC 11179-2:2000, Information technology — Specification and standardization of data elements — Part 2: Classification for data elements
ISOIEC33 (2003) ISO/IEC 11179-3:2003, Information technology — Metadata registries (MDR) — Part 3: Registry metamodel and basic attributes
ISOIEC34 (1996) ISO/IEC 11404:1996, Information technology — Programming languages, their environments and system software interfaces — Language-independent datatypes
ISOIEC35 (1993) ISO/IEC 11544:1993, Information technology — Coded representation of picture and audio information — Progressive bi-level image compression
ISOIEC36 (1995) ISO/IEC 12087-1:1995, Information technology — Computer graphics and image processing — Image processing and interchange (IPI) — Functional specification — Part 1: Common architecture for imaging
ISOIEC37 (1994) ISO/IEC 12087-2:1994, Information technology — Computer graphics and image processing — Image processing and interchange (IPI) — Functional specification — Part 2: Programmer's imaging kernel system application programme interface
ISOIEC38 (1995) ISO/IEC 12087-3:1995, Information technology — Computer graphics and image processing — Image processing and interchange (IPI) — Functional specification — Part 3: Image Interchange Facility (IPI-IIF)

ISOIEC39 (1998) ISO/IEC DIS 12087-5:1998, Information technology — Computer graphics and image processing — Image processing and interchange (IPI) — Functional specification — Part 5: Basic Image Interchange Format (BIIF)
ISOIEC40 (1999) ISO/IEC 13249-3:1999, Information technology — Database languages — SQL Multimedia and Application Packages — Part 3: Spatial, ISO/IEC JTC1 SC 32
ISOIEC41 (1998) ISO/IEC CD 13249-5:1998, Information technology — Database Languages — SQL Multimedia and Application Packages — Part 5: Still Image
ISOIEC42 (1997) ISO/IEC 13522:1997, Information technology — Coding of multimedia and hypermedia information
ISOIEC43 (1996) ISO/IEC 13818:1996, Information technology — Generic coding of moving pictures and associated audio information
ISOIEC44 (1996) ISO/IEC 14252:1996 Information technology — Guide to the POSIX Open System Environment (OSE)
ISOIEC45 (2003) ISO/IEC CD 14481:—, Information technology — Conceptual Schema Modelling Facilities (CSMF)
ISOIEC46 (1999) ISO/IEC 14750:1999, Information technology — Open Distributed Processing — Interface Definition Language, (Formerly DIS 14750-1)
ISOIEC47 (2003) ISO/IEC DIS 19501-1:—, Information technology — Unified Modelling Language (UML) — Part 1: Specification
ITU (2003) http://www.itu.int
Jech T (2002) Set Theory. The Stanford Encyclopedia of Philosophy (Fall 2002 edn). In: Zalta EN (ed), <http://plato.stanford.edu/archives/fall2002/entries/set-theory/>
JTC1 (2003) http://www.jtc1.org
Kim TJ et al. (2002) Workshop for GIS standards: Understanding and Implementation. University of Illinois at Urbana Champaign
Knickmeyer ET (2003) Ellipsoidal and projected coordinate reference system. Personal communication, FH Neubrandenburg, Germany
Lalonde W (ed.) (2002) Styled Layer Descriptor Implementation Specification. OGC Document 02-070, Wayland, Massachusetts
Larisch HJ (2003) Clothoid. Personal communication, FH Neubrandenburg, Germany
Laurini R, Thompson D (1992) Fundamentals of Spatial Information Systems, Academic Press
Lindig G (ed.) (1993) Deutsches Fachwörterbuch Photogrammetrie und Fernerkundung, Part of ISPRS Multilingual Dictionary, Issue ISSN 0344-5879, IfAG (today BKG), Frankfurt, Germany
Longley PA, Goodchild MF, Maguire DJ, Rhind DW (2001) Geographic Information, Systems and Science. Wiley, Hoboken, New Jersey
Luhmann T (2000) Nahbereichsphotogrammetrie (close range photogrammetry). Wichmann, Heidelberg
Mak R (2002) Java Number Cruncher: The Java Programmer's Guide to Numerical Computing. Prentice Hall, Englewood Cliffs, New Jersey
Moritz H (1980) Geodetic Reference System 1980 (GRS80). Bulletin Géodésique, vol 54 (3): 395-405
NIMA97 (1997) Department of Defense World Geodetic System 1984 – Its Definition and Relationship with Local Geodetic Systems. National Imagery and Mapping Agency (NIMA) Technical Report 8350.2: Third edn, <http://164.214.2.59/GandG/tr8350_2.html>
NIMA90 (1990) Datums, Ellipsoids, Grids and Grid Reference Systems. National Imagery and Mapping Agency (NIMA) Technical Report 8358.1, <http://164.214.2.59/GandG/tm83581/toc.htm>

Nitschke M (2003) Mathematical curves. Personal communication, designed with MATLAB, FH Neubrandenburg, Germany

OED (1970) The Oxford English Dictionary, Oxford University Press, Oxford

OGC (2003) http://www.opengis.org

OGC96 (1996) The OpenGIS Abstract Specification: An Object Model for Interoperable Geoprocessing, Revision 1. OpenGIS Consortium (OGC), Document 96-015R1, Wayland, Massachusetts, <http://www.opengis.org>

OGC98a (1998) The OpenGIS Abstract Specification, Volumes 1-16, OpenGIS Consortium (OGC), Wayland, Massachusetts, <http://www.opengis.org>

OGC98b(1998) The OpenGIS Abstract Specification, Topic 7: The Earth Imagery Case, Version 3. OpenGIS Consortium (OGC), Document 98-107, Wayland, Massachusetts, <http://www.opengis.org>

OGC00 (2000) The OpenGIS Abstract Specification, Topic 6: The Coverage Type and its Subtypes. OpenGIS Consortium (OGC), Document 00-106, Wayland, Massachusetts, <http://www.opengis.org>

OMG (2003) UML Resource Page. Object Management Group, <http://www.omg.org/uml>

Parise F (1982) The book of calendars. Facts on file, New York

Portele C (2003) Coordinate Reference Systems in GML. Personal communication, Interactive Instruments, Bonn, Germany

POSC (2003) Specifications and Recommendations. Petrotechnical Open Standards Consortium (POSC), <http://www.posc.org/>

RDF99 (1999) Resource Description Framework(RDF), Model and Syntax Specification. W3C Recommendation 22 February 1999, <http://www.w3.org/TR/REC-rdf-syntax>

RDF00 (2000) Resource Description Framework(RDF), Schema Specification 1.0. W3C Candidate Recommendation 27 March 2000, <http://www.w3.org/TR/2000/CR-rdf-schema-20000327>

Ritter N, Ruth M (1997) The GeoTIFF Data Interchange Standard for Raster Geographic Images. International Journal for Remote Sensing, vol 18 (7): 1637-1647, Taylor & Francis, London

Rugg, RD, Egenhofer MJ, Kuhn W (1995) Formalizing Behavior of Geographic Feature Types. National Center for Geographic Information and Analysis, Report 95-7, Geographical systems, Springer, New York

Rumbaugh J, Booch G, Jacobsson I (1999) UML Reference Manual. Addison-Wesley, Boston

S57 (1996) IHO Transfer Standard for Digital Hydrographic Data, S-57. International Hydrographic Organization, edn 3.0, Monaco

Schulz F (1995) Architectural Framework for Information Infrastructure. Open System Environment (OSE), National Institute of Standards and Technology (NIST) Special Publication 500-232, Gaithersburg, Maryland

SIA (2003) http://www.standardsinaction.org/tc211terms

SMIL (2001) Synchronized Multimedia Integration Language (SMIL 2.0). W3C Recommendation, 07 August 2001, <http://www.w3.org./TR/smil20/>

STANAG98 (2000) Standardization Agreement (STANAG) 4545 –Secondary Imagery Format (NSIF). NATO Allied Engineering Documentation Publication, edition 1, amendment 1

STANAG97 (1997) Standardization Agreement (STANAG) 7074 – Digital Geographic Information Exchange Standard (DIGEST). NATO Allied Geographic Publication (AGeoP) 3

Suh S, Park C, Kim TJ (1990) A Highway Capacity Function in Korea: Measurement and Calibration. Transportation Research, 24A(3):176-186

SVG (2000) Scalable Vector Graphics (SVG) 1.0 Specification. W3C Candidate Recommendation, 02 November 2000, <http://www.w3.org/TR/SVG/>

Varma H (2003) Riemann hyperspace. Personal communication, Canadian Hydrographic Service, Halifax, Canada

Warmer J, Kleppe AG (1997) The Object Constraint Language: Precise modelling with UML. Addison-Wesley, Boston

Worboys MF (1992) A generic model for planar geographical objects. International Journal of Geographical Information Systems, vol 6 (5): 353-372, Taylor & Francis, London

Worboys MF and Bofakos P (1993) A Canonical model for a class of areal spatial objects. Advances in Spatial Databases — Third International Symposium, SSD '93, vol 692: 36-52, Lecture Notes in Computer Science, Springer, Singapore

XML99 (1999) Namespaces in XML, W3C Recommendation, 14 January 1999, <http://www.w3.org/TR/1999/REC-xml-names-19990114>

XML00 (2000) Linking Language (XLink) Version 1.0. W3C Proposed Recommendation, 20 December 2000, <http://www.w3.org/TR/2000/PR-xlink-20001220/>

XML01 (2001) Pointer Language (XPointer) Version 1.0. W3C Last Call Working Draft 8 January 2001, <http://www.w3.org/TR/xptr>

XMI03 (2003) XML Metadata Interchange (XMI) version 2.0. Object Management Group, <http://www.omg.org/technology/documents/formal/xmi.htm>

XSL (1999) Extensible Stylesheet Language Transformations (XSLT), Version 1.0. W3C Recommendation, 16 November 1999,
<http://www.w3.org/TR/1999/REC-xslt-19991116>

You J, Kim TJ (2000) Development and Evaluation of A Hybrid Travel Time Forecasting Model. In: Thill JC (ed), Geographic Information Systems in Transportation Research, Elsevier Science, Oxford

Zhan FB, Noon CE (1996) Shortest Path Algorithms: An Evaluation Using Real Road Networks. Transportation Science, 32(1):65-73, Kluwer Academic, Boston

Zhan FB (1997) Three Fastest Shortest Path Algorithms on Real Road Networks: Data Structures and Procedures. Journal of Geographic Information and Decision Analysis, Volume 1, No 1, University of Western Ontario, London, Ontario, <http://publish.uwo.ca/~jmalczew/gida_1/Zhan/Zhan.htm>

Zhan FB, Noon CE (2000) A Comparison Between Label-Setting and Label-Correcting Algorithms for Computing One-to-One Shortest Paths. In Journal of Geographic Information and Decision Analysis, Volume 4, No 2, University of Western Ontario, London, Ontario

1)
CD = Committee Draft
DIS = Draft International Standard
PDTS = Proposed Draft Technical Specification
WD = Working Draft

Annex A: Terms and definitions of the ISO 19100 standards

This listing of terms and definitions has been compiled from the ISO 19100 standards that have been published by June 2003 or that have reached the enquiry or the approval stage (DIS or FDIS).

Every definition is followed by a source remark pointing to the standard, in which the term has originally been defined. If the definition has been adopted from standards other than the geomatics standards then the ISO 19100 standard is mentioned as well.

A few of the terms still have inconsistent definitions such as "spatial object". In those cases both definitions are given. The harmonization of all terms is ongoing in the ISO/TC211 groups.

abbreviation
designation formed by omitting words or letters from a longer form and designating the same concept (ISO10 2000, ISO36 2002)

abstract test case
generalized test for a particular requirement (ISO37 2000)
NOTE   An abstract test case is a formal basis for deriving test cases. One or more test purposes are encapsulated in the abstract test case. An abstract test case is independent of both the implementation and the values. It should be complete in the sense that it is sufficient to enable a test verdict to be assigned unambiguously to each potentially observable test outcome (i.e. sequence of test events).

abstract test method
method for testing of implementations independent of any particular test procedures (ISO37 2000)

abstract test module
set of related abstract test cases (ISO37 2000)
NOTE   Abstract test modules may be nested in a hierarchical way.

abstract test suite [ATS]
abstract test module specifying all the requirements to be satisfied conformance (ISO37 2000)
NOTE   Abstract test suites are described in a conformance clause.

accuracy
closeness of agreement between a test result and the accepted reference value (ISO45 2002, ISO48 2002)
NOTE 1   A test result can be observations or measurements.
NOTE 2   For positioning services, the test result is a measured value or set of values.

admitted term
term rated according to the scale of the term acceptability rating as a synonym for a preferred term (ISO10 2000, ISO36 2002)

annotation
any marking on illustrative material for the purpose of clarification, such as numbers, letters, symbols, and signs (ISO49 2002)

application
manipulation and processing of data in support of user requirements (ISO34 2002)

Annex A: Terms and definitions of the ISO 19100 standards 227

application schema
conceptual schema for data required by one or more applications (ISO34 2002)
NOTE   An application schema describes the content, the structure and the constraints applicable to information in a specific application domain.

attitude
orientation of a body, described by the angles between the axes of that body's coordinate system and the axes of an external coordinate system (ISO48 2002)
NOTE   In positioning services, this is usually the orientation of the user's platform, such as an aircraft, boat, or automobile.

bag
finite, unordered collection of related items (objects or values) that may be repeated (ISO39 2003)
NOTE   Logically, a bag is a set of pairs <item, count>.

base standard
ISO geographic information standard or other Information Technology standards that can be used as a source from which a profile may be constructed (ISO34 2002)

basic test
initial capability test intended to identify clear cases of non-conformance (ISO37 2000)

boundary
set that represents the limit of an entity (ISO39 2003)
NOTE   Boundary is most commonly used in the context of geometry, where the set is a collection of points or a collection of objects that represent those points. In other arenas, the term is used metaphorically to describe the transition between an entity and the rest of its domain of discourse.

buffer
geometric object that contains all direct positions whose distance from a specified geometric object is less than or equal to a given distance (ISO39 2003)

calendar
discrete temporal reference system that provides a basis for defining temporal position to a resolution of one day (ISO40 2002)
calendar era
sequence of periods of one of the types used in a calendar, counted from a specified event (ISO40 2002)

capability test
test designed to determine whether an IUT conforms to a particular characteristic of a standard as described in the test purpose (ISO37 2000)

Cartesian coordinate system
coordinate system which gives the position of points relative to N mutually-perpendicular axes(ISO43 2003)
NOTE N is 1, 2 or 3 for the purposes of ISO 19111.

character
member of a set of elements that is used for the representation, organization, or control of data (ISOIEC07 1993, ISO50 2002)

circular sequence
sequence which has no logical beginning and is therefore equivalent to any circular shift of itself; hence the last item in the sequence is considered to precede the first item in the sequence (ISO39 2003)

class
description of a set of objects that share the same attributes, operations, methods, relationships, and semantics (ISO39 2003)
NOTE   A class may use a set of interfaces to specify collections of operations it provides to its environment. The term was first used in this way in the general theory of object oriented programming, and later adopted for use in this same sense in UML.

closure
union of the interior and boundary of a topological or geometric object (ISO39 2003)

coboundary
set of topological primitives of higher topological dimension associated with a particular topological object, such that this topological object is in each of their boundaries (ISO39 2003)
NOTE   If a node is on the boundary of an edge, that edge is on the coboundary of that node. Any orientation parameter associated to one of these relations would also be associated to the other. So that if the node is the end node of the edge (defined as the end of the positive directed edge), then the positive orientation of the node (defined as the positive directed node) would have the edge on its coboundary.

code
representation of a label according to a specified scheme (ISO50 2002)

complex feature
feature composed of other features (ISO41 2002)

composite curve
sequence of curves such that each curve (except the first) starts at the end point of the previous curve in the sequence (ISO39 2003)

NOTE   A composite curve, as a set of direct positions, has all the properties of a curve.

composite solid
connected set of solids adjoining one another along shared boundary surfaces (ISO39 2003)
NOTE   A composite solid, as a set of direct positions, has all the properties of a solid.

composite surface
connected set of surfaces adjoining one another along shared boundary curves (ISO39 2003)
NOTE   A composite surface, as a set of direct positions, has all the properties of a surface.

compound Coordinate Reference System
description of position through two independent Coordinate Reference Systems to describe a position (ISO43 2003)
EXAMPLE   One Coordinate Reference System based on a two- or three-dimensional coordinate system and the other Coordinate Reference System based on a gravity-related height system.

computational geometry
manipulation of and calculations with geometric representations for the implementation of geometric operations (ISO39 2003)
EXAMPLE   Computational geometry operations include testing for geometric inclusion or intersection, the calculation of convex hulls or buffer zones, or the finding of shortest distances between geometric objects.

computational topology
topological concepts, structures and algebra that aid, enhance or define operations on topological objects usually performed in computational geometry (ISO39 2003)

concept
unit of knowledge created by a unique combination of characteristics (ISO10 2000, ISO36 2002)
NOTE   Concepts are not necessarily bound to particular languages. They are, however, influenced by the social or cultural background which often leads to different categorizations.

concept system
set of concepts structured according to the relations among them (ISO10 2000, ISO36 2002)

concept harmonization

activity for reducing or eliminating minor differences between two or more concepts which are already closely related to each other (ISO09 1996, ISO36 2002)

conceptual formalism
set of modelling concepts used to describe a conceptual model (ISO34 2002)
EXAMPLE   UML meta model, EXPRESS meta model
NOTE   One conceptual formalism can be expressed in several conceptual schema languages.

conceptual model
model that defines concepts of a universe of discourse (ISO34 2002)
NOTE   Well defined means that the definition is both necessary and sufficient, as everything that satisfies the definition is in the set and everything that does not satisfy the definition is necessarily outside the set.

conceptual schema
formal schema of a conceptual model. (ISO34 2002)

conceptual schema language
formal language based on a conceptual formalism for the purpose of representing conceptual schemas (ISO34 2002)
EXAMPLE   UML, EXPRESS, IDEF1X
NOTE   A conceptual schema language may be lexical or graphical. Several conceptual schema languages can be based on the same conceptual formalism.

conformance
fulfilment of specified requirements (ISO37 2000)

conformance assessment process
process for assessing the conformance of an implementation to a standard (ISO37 2000)

conformance clause
clause defining what is necessary in order to meet the requirements of the standard (ISO37 2000)

conformance quality level
threshold value or set of threshold values for data quality results used to determine how well a dataset meets the criteria set forth in its product specification or user requirements (ISO46 2001)

conformance testing
testing of a product to determine the extent to which the product is a conforming implementation (ISO37 2000)

conformance test report

summary of the conformance to the standard as well as all the details of the testing that supports the given overall summary (ISO37 2000)

conforming implementation
implementation which satisfies the requirements (ISO37 2000)

connected
property of a geometric object implying that any two direct positions on the object can be placed on a curve that remains totally within the object (ISO39 2003)
NOTE   A topological object is connected if and only if all its geometric realizations are connected. This is not included as a definition because it follows from a theorem of topology.

connected node
node that starts or ends one or more edges (ISO39 2003)

conversion rule
rule for converting instances in the input data structure to instances in the output data structure (ISO50 2002)

convex hull
smallest convex set containing a given geometric object (Illingworth 1997, ISO39 2003)
NOTE   "Smallest" is the set theoretic smallest, not an indication of a measurement. The definition can be rewritten as "the intersection of all convex sets that contain the geometric object".

convex set
geometric set in which any direct position on the straight-line segment joining any two direct positions in the geometric set is also contained in the geometric set (Illingworth 1997, ISO39 2003)
NOTE   Convex sets are "simply connected", meaning that they have no interior holes, and can normally be considered topologically isomorphic to a Euclidean ball of the appropriate dimension. So the surface of a sphere can be considered to be geodesically convex.

coordinate
one of a sequence of N numbers designating the position of a point in N-dimensional space (ISO43 2003)
NOTE   In a Coordinate Reference System, the numbers must be qualified by units.

coordinate conversion
change of coordinates, based on a one-to-one relationship, from one coordinate system to another based on the same datum (ISO43 2003)

EXAMPLE   Between geodetic and Cartesian coordinate systems or between geodetic coordinates and projected coordinates, or change of units such as from radians to degrees or feet to metres.
NOTE   A coordinate conversion uses parameters which have constant values.

coordinate dimension
number of measurements or axes needed to describe a position in a coordinate system (ISO39 2003)

coordinate operation
change of coordinates, based on a one-to-one relationship, from one Coordinate Reference System to another (ISO43 2003)
NOTE   Supertype of coordinate transformation and coordinate conversion

Coordinate Reference System
coordinate system that is related to the real world by a datum (ISO43 2003)
NOTE   For geodetic and vertical datums, it will be related to the Earth.

coordinate system
set of mathematical rules for specifying how coordinates are to be assigned to points (ISO43 2003)

coordinate transformation
change of coordinates from one Coordinate Reference System to another Coordinate Reference System based on a different datum through a one-to-one relationship (ISO43 2003)
NOTE   A coordinate transformation uses parameters which may are derived empirically by a set of points with known coordinates in both Coordinate Reference Systems.

Coordinated Universal Time (UTC)
time scale maintained by the Bureau International des Poids et Mesures (International Bureau of Weights and Measures) and the International Earth Rotation Service (IERS) that forms the basis of a coordinated dissemination of standard frequencies and time signals (ISO40 2002)

curve
1-dimensional geometric primitive, representing the continuous image of a line (ISO39 2003)
NOTE   The boundary of a curve is the set of points at either end of the curve. If the curve is a cycle, the two ends are identical, and the curve (if topologically closed) is considered to not have a boundary. The first point is called the start point, and the last is the end point. Connectivity of the curve is guaranteed by the "continuous image of a line" clause. A topological theorem states that a continuous image of a connected set is connected.

curve segment
1-dimensional geometric object used to represent a continuous component of a curve using homogeneous interpolation and definition methods (ISO39 2003)
NOTE   The geometric set represented by a single curve segment is equivalent to a curve.

cycle
<geometry>
spatial object without a boundary (ISO39 2003)
NOTE   Cycles are used to describe boundary components (see shell, ring). A cycle has no boundary because it closes on itself, but it is bounded (i.e., it does not have infinite extent). A circle or a sphere, for example, has no boundary, but is bounded.

data
reinterpretable representation of information in a formalised manner suitable for communication, interpretation, or processing (ISOIEC07 1993, ISO50 2002)

data element
unit of data that, in a certain context, is considered indivisible (ISO50 2002)

data interchange
delivery, receipt and interpretation of data (ISO50 2002)

dataset
identifiable collection of data (ISO47 2003)

data level
level containing data describing specific instances (ISO34 2002)

data quality date
date or range of dates on which a data quality measure is applied (ISO45 2002)

data quality element
quantitative component documenting the quality of a dataset
NOTE   The applicability of a data quality element to a dataset depends on both the dataset's content and its product specification; the result being that all data elements may not be applicable to all datasets. (ISO34 2002)

data quality element
quantitative component documenting the quality of a dataset (ISO45 2002)
NOTE   The applicability of a data quality element to a dataset depends on both the dataset's content and its product specification; The result being that all data quality elements may not be applicable to all datasets.

data quality evaluation procedure
operation(s) used in applying and reporting quality evaluation methods and their results (ISO45 2002)

data quality measure
evaluation of a data quality subelement (ISO45 2002)
EXAMPLE   The percentage of the values of an attribute that are correct.

data quality overview element
non-quantitative component documenting the quality of a datasets. (ISO45 2002)
NOTE   Information about the purpose, usage, and lineage of a dataset is non-quantitative information.

data quality result
value or set of values resulting from applying a data quality measure or the outcome of evaluating the obtained value or set of values against a specified acceptable quality level (ISO45 2002)
EXAMPLE   A data quality result of "90" with a data quality value type of "percentage" reported for the data quality element and its data quality subelement "completeness/commission" is an example of a value resulting from applying a data quality measure to the data specified by a data quality scope. A data quality result of "true" with a data quality value type of "boolean variable" is an example of comparing the value (90) against a specified acceptable conformance quality level (85) and reporting an evaluation of a kind, pass or fail.

data quality scope
extent or characteristic(s) of the data for which quality information is reported (ISO45 2002)
NOTE   A data quality scope for a dataset can comprise a dataset series to which the dataset belongs, the dataset itself, or a smaller grouping of data located physically within the dataset sharing common characteristics. Common characteristics can be an identified feature type, feature attribute, or feature relationship; data collection criteria; original source; or a specified geographic or temporal extent.

data quality subelement
component of a data quality element describing a certain aspect of that data quality element (ISO45 2002)

data quality value type
value type for reporting a data quality result (ISO45 2002)
EXAMPLES   "boolean variable", "percentage", "ratio"
NOTE   A data quality value type is always provided for a data quality result.

data quality value unit
value unit for reporting a data quality result. (ISO45 2002)

EXAMPLE "metre"
NOTE A data quality value unit is provided only when applicable for a data quality result.

data transfer
movement of data from one point to another over a medium (ISO50 2002)
NOTE Transfer of information implies transfer of data.

data type
specification of the legal value domain and legal operations allowed on values in this domain
EXAMPLE Integer, Real, Boolean, String, Date, and GM_Point.
NOTE A data type is identified by a term, e.g. Integer
NOTE A data type is identified by a term, e.g. Integer. Values of the data types shall be of the specified value domain, e.g. all integer numbers between − 65537 and 65536. The set of operations can be +, -, / and * and shall have a well defined semantic. A data type can be simple or complex. A simple data type defines a value domain where values are considered atomic in a certain context, e.g. Integer. A complex data type is a collection of data types, which are grouped together. A complex data type may represent an object and can thus have identity.

dataset
identifiable collection of data (ISO34 2002)
NOTE A dataset may be a smaller grouping of data which, though limited by some constraint such as spatial extent or feature type, is located physically within a larger dataset. Theoretically, a dataset may be as small as a single feature or feature attribute contained within a larger dataset. A hardcopy map or chart may be considered a dataset.
NOTE For purposes of data quality evaluation, a dataset may be as small as a single feature or feature attribute contained within a larger dataset.
NOTE The principles which apply to datasets may also be applied to dataset series and reporting groups.

dataset series
collection of datasets sharing the same product specification (ISO47 2003)

datum
parameter or set of parameters that serve as a reference or basis for the calculation of other parameters (ISO43 2003)
NOTE 1 A datum defines the position of the origin, the scale, and the orientation of the axes of a coordinate system.
NOTE 2 A datum may be a geodetic datum, a vertical datum or an engineering datum.

day

period having a duration nominally equivalent to the periodic time of the Earth's rotation around its axis

definition
representation of a concept by a descriptive statement which serves to differentiate it from related concepts (ISO10 2000, ISO36 2002)

deprecated term
term rated according to the scale of the term acceptability rating as undesired (ISO10 2000, ISO36 2002)

direct evaluation method
method of evaluating the quality of a dataset based on inspection of the items within the dataset (ISO46 2001)

direct position
position described by a single set of coordinates within a Coordinate Reference System (ISO39 2003)

directed edge
directed topological object that represents an association between an edge and one of its orientations (ISO39 2003)
NOTE   A directed edge that is in agreement with the orientation of the edge has a + orientation, otherwise, it has the opposite (-) orientation. Directed edge is used in topology to distinguish the right side (-) from the left side (+) of the same edge and the start node (-) and end node (+) of the same edge and in computational topology to represent these concepts.

directed face
directed topological object that represents an association between a face and one of its orientations (ISO39 2003)
NOTE   The orientation of the directed edges that compose the exterior boundary of a directed face will appear positive from the direction of this vector; the orientation of a directed face that bounds a topological solid will point away from the topological solid. Adjacent solids would use different orientations for their shared boundary, consistent with the same sort of association between adjacent faces and their shared edges. Directed faces are used in the coboundary relation to maintain the spatial association between face and edge.

directed node
directed topological object that represents an association between a node and one of its orientations (ISO39 2003)
NOTE   Directed nodes are used in the coboundary relation to maintain the spatial association between edge and node. The orientation of a node is with respect to an edge, "+" for end node, "-" for start node. This is consistent with the vector notion of "result = end – start".

directed solid
directed topological object that represents an association between a topological solid and one of its orientations (ISO39 2003)
NOTE   Directed solids are used in the coboundary relation to maintain the spatial association between face and topological solid. The orientation of a solid is with respect to a face, "+" if the upNormal is outward, "-" if inward. This is consistent with the concept of "up = outward" for a surface bounding a solid.

directed topological object
topological object that represents a logical association between a topological primitive and one of its orientations (ISO39 2003)

domain
well-defined set (ISO35 2001)
NOTE   Domains are used to define the domain and range of operators and functions.

easting (*E*)
distance in a coordinate system, eastwards (positive) or westwards (negative) from a north-south reference line (ISO43 2003)

edge
1-dimensional topological primitive (ISO39 2003)
NOTE   The geometric realization of an edge is a curve. The boundary of an edge is the set of one or two nodes associated to the edge within a topological complex.

edge-node graph
graph embedded within a topological complex composed of all of the edges and connected nodes within that complex (ISO39 2003)
NOTE   The edge-node graph is a subcomplex of the complex within which it is embedded.

ellipsoid
surface formed by the rotation of an ellipse about a main axis (ISO43 2003)
NOTE   In ISO 19111, ellipsoids are always oblate, meaning that the axis of rotation is always the minor axis.
NOTE   Only used as part of a three-dimensional geodetic coordinate system and never on its own.

ellipsoidal coordinate system
coordinate system in which position is specified by geodetic latitude, geodetic longitude and (in the three-dimensional case) ellipsoidal height (ISO43 2003)
geodetic datum
datum describing the relationship of a coordinate system to the Earth (ISO43 2003)

NOTE   In most cases, the geodetic datum include an ellipsoid definition.

ellipsoidal height, geodetic height *h*
distance of a point from the ellipsoid measured along the perpendicular from the ellipsoid to this point positive if upwards or outside of the ellipsoid (ISO43 2003)
NOTE   Only used as part of a three-dimensional geodetic coordinate system and never on its own.

encapsulation
collection of specified data content in a well-defined coding structure or the process by which it is done (ISO52 2001)

encoding
conversion of data into a series of codes (ISO50 2002)

encoding rule
identifiable collection of conversion rules that define the encoding for a particular data structure (ISO50 2002)
EXAMPLE   XML, ISO/IEC 8211.
NOTE   An encoding rule specifies the types of data to be converted as well as the syntax, structure and codes used in the resulting data structure.

encoding service
software component that has an encoding rule implemented (ISO50 2002)

end node
node in the boundary of an edge that corresponds to the end point of that edge as a curve in any valid
geometric realization of a topological complex in which the edge is used (ISO39 2003)

end point
last point of a curve (ISO39 2003)

engineering datum
local datum
datum describing the relationship of a coordinate system to a local reference (ISO43 2003)
NOTE   Engineering datum excludes both geodetic and vertical datums.
EXAMPLE   A system for identifying relative positions within a few kilometres of the reference point.

event
action which occurs at an instant (ISO40 2002)

executable test case

Annex A: Terms and definitions of the ISO 19100 standards

specific test of an implementation to meet particular requirements (ISO37 2000)
NOTE   Instantiation of an abstract test case with values

executable test suite (ETS)
set of executable test cases (ISO37 2000)

exterior
difference between the universe and the closure (ISO39 2003)
NOTE   The concept of exterior is applicable to both topological and geometric complexes.

external function
function not part of the application schema (ISO49 2002)
NOTE   The electronic map in a car navigation system has to be displayed so that the up-direction of the map is always in the direction the car is moving. To be able to specify the rotation of the map, the current position of the car must be retrieved continuously from an external position device using an external function.

face
2-dimensional topological primitive (ISO39 2003)
NOTE   The geometric realization of a face is a surface. The boundary of a face is the set of directed edges within the same topological complex that are associated to the face via the boundary relations. These can be organized as rings.

fail verdict
test verdict reporting of non-conformance (ISO37 2000)
NOTE   Non-conformance may be with respect to either the test purpose or at least one of the conformance requirements of the relevant standard(s).

falsification test
test to find errors in the implementation (ISO37 2000)
NOTE   If errors are found, one can correctly deduce the implementation does not conform to the standard; however, the absence of errors does not necessarily imply the converse. Falsification test can only demonstrate non-conformance. Compare with verification test. Due to technical and economical problems, in most cases, falsification test is adopted as a test method for conformance testing.

feature
abstraction of real world phenomena (ISO34 2002)
NOTE 1   A feature may occur as a type or an instance. Feature type or feature instance shall be used when only one is meant.
NOTE 2   In a feature catalogue, the basic level of classification is the feature type
EXAMPLE   The phenomenon "Eiffel Tower" may be classified with other similar phenomena into a feature type "tower".

feature association
relationship that links instances of one feature type with instances of the same or a different feature type (ISOIEC33 2003, ISO42 2001)
NOTE 1   A feature association may occur as a type or an instance. Feature association type or feature association instance is used when only one is meant.
NOTE 2   Feature associations include aggregation of features.

feature attribute
characteristic of a feature (ISO34 2002)
EXAMPLE 1   A feature attribute named "colour" may have an attribute value "green" which belongs to the data type "text".
EXAMPLE 2   A feature attribute named "length" may have an attribute value "82.4" which belongs to the data type "real".
NOTE 1   A feature attribute has a name, a data type, and a value domain associated to it. A feature attribute for a feature instance also has an attribute value taken from the value domain.
NOTE 2   In a feature catalogue, a feature may include a value domain but does not specify attribute values for feature instances.

feature catalogue
catalogue containing definitions and descriptions of the features types, feature attributes, and feature relationships, occurring in one or more sets of geographic data, together with any feature operations that may be applied (ISO34 2002)

feature division
feature succession in which a previously existing feature is replaced by two or more distinct feature instances of the same feature type (ISO40 2002)
EXAMPLE   An instance of the feature type "land parcel" is replaced by two instances of the same type when the parcel is legally subdivided.

feature fusion
feature succession in which two or more previously existing instances of a feature type are replaced by a single instance of the same feature type (ISO40 2002)
EXAMPLE   Two instances of the feature type "pasture" are replaced by a single instance when the fence between the pastures is removed.

feature operation
operation that every instance of a feature type may perform (ISO34 2002)
EXAMPLE 1   An operation upon the feature type "dam" is to raise the dam. The result of this operation is to raise the level of water in a reservoir.
EXAMPLE 2   An operation by the feature type "dam" might be to block vessels from navigating along a watercourse.
NOTE   Feature operations provide a basis for feature type definition.

feature portrayal rule set
collection of portrayal rules that apply to a feature instance (ISO49 2002)

feature substitution
feature succession in which one feature instance is replaced by another feature instance of the same or different feature type (ISO40 2002)
EXAMPLE   An instance of feature "type building" is razed and replaced by an instance of feature type "parking lot".

feature succession
replacement of one or more feature instances by other feature instances, such that the first feature instances cease to exist (ISO40 2002)

feature table
a table where the columns represent feature attributes, and the rows represent features (ISO55 2000)

file
named set of records stored or processed as a unit (ISOIEC07 1993, ISO50 2002)

flattening (f)
ratio of the difference between the semi-major (a) and semi-minor axis (b) of an ellipsoid to the semi-major axis; $f = (a-b)/a$ (ISO43 2003)
NOTE   Sometimes inverse flattening $1/f = a/(a-b)$ is given instead; $1/f$ is also known as reciprocal flattening.

full inspection
inspection of every item in a dataset (ISO14 1993, ISO46 2001)
NOTE   Full inspection is also known as 100% inspection.

function
rule that associates each element from a domain (source, or domain of the function) to a unique element in another domain (target, co-domain, or range) (ISO39 2003)

functional language
programming language in which abstract data types are defined in terms of operations on the types, and in which algebraic axioms specify the results of each of the operations for each of the types (ISO42 2001)
NOTE   In a functional language, feature types may be represented as abstract data types.

functional standard
existing geographic information exchange standard developed specifically for transfer of data between entities in different nations, and currently used for that purpose (ISO52 2001)
NOTE   GDF, S-57, and DIGEST are examples of functional standards

gazetteer

directory of instances of a class or classes of features containing some information regarding position (ISO44 2001)
NOTE   The positional information need not be coordinates, but could be descriptive.
geodetic coordinate system

geodetic latitude (ellipsoidal latitude, φ)
angle from the equatorial plane to the perpendicular to the ellipsoid through a given point, northwards treated as positive (ISO43 2003)

geodetic longitude (ellipsoidal longitude, λ)
angle from the prime meridian plane to the meridian plane of the given point, eastward treated as positive (ISO43 2003)

geographic data
data with implicit or explicit reference to a location relative to the Earth (ISO41 2002)
NOTE   Geographic information is also used as a term for information concerning phenomena implicitly or explicitly associated with a location relative to the Earth.

geographic feature
representation of real world phenomenon associated with a location relative to the Earth (ISO55 2000)

geographic identifier
spatial reference in the form of a label or code that identifies a location (ISO44 2001)
EXAMPLE   "Spain" is an example of a country name, "SW1P 3AD" is an example of a postcode.

geographic information
information concerning phenomena implicitly or explicitly associated with a location relative to the Earth (ISO34 2002)

geographic information service
service that transforms, manages, or presents geographic information to users (ISO34 2002)

geographic information system
information system dealing with information concerning phenomena associated with location relative to the Earth (ISO34 2002)

geoid
level surface which best fits mean sea level either locally or globally (ISO43 2003)
NOTE   "Level surface" means an equipotential surface of the Earth's gravity field which is everywhere perpendicular to the direction of gravity.

geometric aggregate
collection of geometric objects that has no internal structure (ISO39 2003)
NOTE   No assumption about the spatial realtionship between the elements can be made.

geometric boundary
boundary represented by a set of geometric primitives of smaller geometric dimension that limits the extent of a geometric object (ISO39 2003)

geometric complex
set of disjoint geometric primitives where the boundary of each geometric primitive can be represented as the union of other geometric primitives of smaller dimension within the same set (ISO39 2003)
NOTE   The geometric primitives in the set are disjoint in the sense that no direct position is interior to more than one geometric primitive. The set is closed under boundary operations, meaning that for each element in the geometric complex, there is a collection (also a geometric complex) of geometric primitives that represents the boundary of that element. Recall that the boundary of a point (the only 0D primitive object type in geometry) is empty. Thus, if the largest dimension geometric primitive is a solid (3D), the composition of the boundary operator in this definition terminates after at most 3 steps. It is also the case that the boundary of any object is a cycle.

geometric dimension
largest number n such that each direct position in a geometric set can be associated with a subset that has the direct position in its interior and is similar (isomorphic) to R^n, Euclidean n-space (ISO39 2003)
NOTE   Curves, because they are continuous images of a portion of the real line, have geometric dimension 1. Surfaces cannot be mapped to R^2 in their entirety, but around each point position, a small neighbourhood can be found that resembles (under continuous functions) the interior of the unit circle in R^2, and are therefore 2-dimensional. In this standard, most surface patches (instances of GM_SurfacePatch) are mapped to portions of R^2 by their defining interpolation mechanisms.

geometric object
spatial object representing a geometric set (ISO39 2003)
NOTE   A geometric object consists of a geometric primitive, a collection of geometric primitives, or a geometric complex treated as a single entity. A geometric object may be the spatial representation of an object such as a feature or a significant part of a feature.

geometric primitive
geometric object representing a single, connected, homogeneous element of space (ISO39 2003)

NOTE   Geometric primitives are non-decomposed objects that present information about geometric

geometric realization
geometric complex whose geometric primitives are in a 1 to 1 correspondence to the topological primitives of a topological complex, such that the boundary relations in the two complexes agree (ISO39 2003)
NOTE   In such a realization the topological primitives are considered to represent the interiors of the corresponding geometric primitives. Composites are closed.

geometric set
set of direct positions (ISO39 2003)
NOTE   This set in most cases is infinite.

graph
set of nodes, some of which are joined by edges (ISO39 2003)
NOTE   In geographic information systems, a graph can have more than one edge joining two nodes, and can have an edge that has the same node at both ends.

graphical language
language whose syntax is expressed in terms of graphical symbols (ISO34 2002)

gravity-related height (*H*)
height dependent on the Earth's gravity field (ISO43 2003)
NOTE   In particular, orthometric height or normal height, which are both approximations of the distance of a point above the mean sea level.

Greenwich meridian
meridian that passing through the position of the Airy Transit Circle at the Royal Observatory Greenwich, United Kingdom (ISO43 2003)
NOTE   Most geodetic datums use the Greenwich meridian as the prime meridian. Its precise position differs slightly between different datums.

Gregorian calendar
calendar in general use; first introduced in 1582 to define a year that more closely approximated the tropical year than the Julian calendar (ISO17 2000, ISO40 2002)
NOTE   The introduction of the Gregorian calendar icluded the cancellation of the accumulated inaccuracies of the Julian year. In the Gregorian calendar, a calendar year is either a common year or a leap year; each year is devided into 12 sequential months.

grid
network composed of two or more sets of curves in which the member of each set intersect the members of the other sets in an algorithmic way (ISO53 2002)

height (h or H)
distance of a point from a chosen reference surface along a line perpendicular to that surface (ISO43 2003)
NOTE 1 See ellipsoidal height and gravity-related height.
NOTE 2 Height of a point outside the surface treated as positive; negative height is also named as depth.

homomorphism
relationship between two domains (such as two complexes) such that there is a structure preserving function from one to the other (ISO39 2003)
NOTE Homomorphisms are distinct from isomorphisms in that no inverse function is required. In an isomorphism, there are essentially two homomorphisms that are functional inverses of one another. Continuous functions are topological homomorphisms because they preserve "topological characteristics". The mapping of topological complexes to their geometric realizations preserves the concept of boundary and is therefore a homomorphism.

identifier
label that uniquely identifies an item or group of items (ISO50 2002)

implementation
realization of a specification (ISO37 2000)
NOTE In the context of ISO geographic information standards, this includes specifications of geographic information services and datasets.

Implementation conformance statement (ICS)
statement of specification options that have been implemented (ISO37 2000)

Implementation eXtra Information for Testing (IXIT)
statement containing all of the information related to the IUT and its corresponding SUT which will enable the testing laboratory to run an appropriate test suite against that IUT (ISO37 2000)
NOTE IXIT typically provides the details on the organization and storage of concepts in the SUT as well as on the means of access to and modification of the SUT.

inconclusive verdict
test verdict when neither a pass verdict nor a fail verdict apply (ISO37 2000)
indirect evaluation method
method of evaluating the quality of a dataset based on external knowledge (ISO46 2001)
NOTE Examples of external knowledge are dataset lineage, such as production method or source data.

inertial positioning system

positioning system employing accelerometers, gyroscopes, and computer as integral components to determine coordinates of points or objects relative to an initial known reference point (ISO48 2002)

information
knowledge concerning objects, such as facts, events, things, processes, or ideas, including concepts, that within a certain context has a particular meaning (ISOIEC07 1993, ISO50 2002)

instance
object that realizes a class (ISO39 2003)
individual entity with its own identity and value (ISO49 2002)
NOTE [Unified Modelling Language version 1.1]. A descriptor specifies the form and behaviour of a set of instances with similar properties.

instance model
representation model for storing data according to an application schema (ISO50 2002)

instant
0-dimensional geometric primitive representing position in time (ISO40 2002)

integrated positioning system
positioning system incorporating two or more positioning technologies (ISO48 2002)
NOTE The measurements produced by each positioning technology in an integrated system may be any of position, motion, or attitude. There may be redundant measurements. When combined, a unified position, motion, or attitude is determined.

interval scale
scale with an arbitrary origin which can be used to describe both ordering of values and distances between values (ISO40 2002)
NOTE Ratios of values measured on an interval scale have no meaning.

interface
named set of operations that characterize the behaviour of an element (ISO37 2000) (ISOIEC47 2003, ISO50 2002)

interior
set of all direct positions that are on a geometric object but which are not on its boundary (ISO39 2003)
NOTE The interior of a topological object is the homomorphic image of the interior of any of its geometric realizations. This is not included as a definition because it follows from a theorem of topology.

interoperability

capability to communicate, execute programs, or transfer data among various functional units in a manner that requires the user to have little or no knowledge of the unique characteristics of those units (ISOIEC07 1993, ISO50 2002)

isolated node
node not related to any edge (ISO39 2003)

isomorphism
relationship between two domains (such as two complexes) such that there are one-to-one, structure-preserving functions from each domain onto the other, and the composition of the two functions, in either order, is the corresponding identity function (ISO39 2003)
NOTE A geometric complex is isomorphic to a topological complex if their elements are in a one-to-one, dimension- and boundary-preserving correspondence to one another.

item
that which can be individually described or considered (ISO15 1989, ISO46 2001)
NOTE An item can be any part of a dataset, such as a feature, feature relationship, feature attribute, or combination of these.

Julian date
Julian day number followed by the decimal fraction of the day elapsed since the preceding noon (ISO40 2002)

Julian day number
number of days elapsed since Greenwich mean noon on 1 January 4713 BC, Julian proleptic calendar (ISO40 2002)

language identifier
information in a terminological entry which indicates the name of a language (ISO10 2000, ISO36 2002)

lexical language
language whose syntax is expressed in terms of symbols defined as character strings (ISO34 2002)

life span
period during which something exists
NOTE Valid-time lifespan is the period during which an object exists in the modelled reality. Transaction-time lifespan is the period during which a database object is current in the database. (ISO40 2002)

linear positioning system
positioning system that measures distance from a reference point along a route (ISO48 2002)

EXAMPLE   An odometer used in conjunction with predefined mile or kilometre origin points along a route provides a linear reference to a position.

linear reference system
reference system that identifies a location by reference to a segment of a linear geographic feature and distance along that segment from a given point (ISO48 2002)
NOTE   Linear reference systems are widely used in transportation, for example highway name and mile or kilometre marker.

location
identifiable geographic place (ISO44 2001)
EXAMPLE   "Eiffel Tower", "Madrid", "California"

map projection
coordinate conversion from a geodetic coordinate system to a plane (ISO43 2003)

mean sea level
average level of the surface of the sea over all stages of tide and seasonal variations (ISO43 2003)
NOTE   Mean sea level in a local context normally means mean sea level for the region calculated from observations at one or more points over a given period of time. Mean sea level in a global context differs from a global geoid by not more than 2 metres.

medium
substance or agency for storing or transmitting data (ISO50 2002)
EXAMPLE   Compact disc, Internet, radio waves, etc.

meridian
intersection of an ellipsoid by a plane containing the semi-minor axis of the ellipsoid (ISO43 2003)
NOTE   This term is often used for the pole-to-pole arc rather than the complete closed figure.

metadata
data about data (ISO47 2003)

metadata element
discrete unit of metadata (ISO47 2003)
NOTE 1   Metadata elements are unique within a metadata entity.
NOTE 2   Equivalent to an attribute in UML terminology.

metadata entity
set of metadata elements describing the same aspect of data (ISO47 2003)
NOTE 1   May contain one or more metadata entities.
NOTE 2   Equivalent to a class in UML terminology.

Annex A: Terms and definitions of the ISO 19100 standards 249

metadata schema
conceptual schema describing metadata (ISO34 2002)
NOTE   ISO 19115 describes a standard for a metadata schema.

metadata section
subset of metadata which consists of a collection of related metadata entities and metadata elements (ISO47 2003)

model
abstraction of some aspects of a universe of discourse (ISO47 2003)
abstraction of some aspects of reality (ISO41 2002)

module
predefined set of elements in a base standard that may be used to construct a profile (ISO52 2001)

month
period approximately equal in duration to the periodic time of a lunar cycle (ISO40 2002)
NOTE   The duration of a month is an integer number of days. The number of days in a month is determined by the rules of the particular calendar.

motion
change in the position of an object over time, represented by change of coordinate values with respect to a particular reference frame (ISO48 2002)
EXAMPLE   This may be motion of the position sensor mounted on a vehicle or other platform or motion of an object being tracked by a positioning system.

neighbourhood
geometric set containing a specified direct position in its interior, and containing all direct positions within a specified distance of the specified direct position (ISO39 2003)

node
0-dimensional topological primitive (ISO39 2003)
NOTE   The boundary of a node is the empty set.

non-conformance
failure to fulfil one or more requirements specified (ISO37 2000)

northing (*N*)
distance in a coordinate system, northwards (positive) or southwards (negative) from an east-west reference line (ISO43 2003)

object

entity with a well defined boundary and identity that encapsulates state and behaviour (ISO39 2003)
NOTE   This term was first used in this way in the general theory of object oriented programming, and later adopted for use in this same sense in UML. An object is an instance of a class. Attributes and relationships represent state. Operations, methods, and state machines represent behaviour.

operating conditions
parameters influencing the determination of coordinate values by a positioning system (ISO48 2002)
NOTE   Measurements acquired in the field are affected by many instrumental and environmental factors, including meteorological conditions, computational methods and constraints, imperfect instrument construction, incomplete instrument adjustment or calibration, and, in the case of optical measuring systems, the personal bias of the observer. Solutions for positions may be affected by the geometric relationships of the observed data and/or mathematical model employed in the processing software.

obsolete term
term which is no longer in common use (ISO10 2000, ISO36 2002)

optical positioning system
positioning system that determines of the position of an object by means of the properties of light (ISO48 2002)
EXAMPLE   Total Station: Commonly used term for an integrated optical positioning system incorporating an electronic theodolite and an electronic distance measuring instrument into a single unit with an internal microprocessor for automatic computations.

ordinal era
one of a set of named periods ordered in time (ISO40 2002)
ordinal temporal reference system
temporal reference system composed of a hierarchy of ordinal eras (ISO40 2002)

ordinal scale
scale which provides a basis for measuring only the relative position of an object (ISO40 2002)

open systems environment (OSE)
comprehensive set of interfaces, services and supporting formats, plus user aspects, for interoperability and/or portability of applications, data, or people, as specified by Information Technology standards and profiles (ISOIEC21 1998, ISO38 2002)

pass verdict
test verdict of conformance (ISO37 2000)

performance indicator
internal parameters of positioning systems indicative of the level of performance achieved (ISO48 2002)
NOTE   Performance indicators can be used as quality control evidence of the positioning system and/or positioning solution. Internal quality control may include such factors as signal strength of received radio signals (signal-to-noise ratio (SNR)), figures indicating the dilution of precision (DOP) due to geometric constraints in radiolocation systems, and system specific figure of merit (FOM).

period
one dimensional geometric primitive representing extent in time (ISO40 2002)
NOTE   A period is bounded by two different temporal positions

periodic time
duration of one cycle (ISO05 1992, ISO40 2002)

planar topological complex
topological complex that has a geometric realization that can be embedded in Euclidean 2 space (ISO39 2003)

point
0-dimensional geometric primitive, representing a position (ISO39 2003)
NOTE   The boundary of a point is the empty set.

polar coordinate system
coordinate system in which position is specified by distance and direction from the origin (ISO43 2003)
NOTE   In three dimensions also called spherical coordinate system.

population
totality of items under consideration (ISO14 1993, ISO46 2001)
EXAMPLE 1   All points in a dataset.
EXAMPLE 2   Names of all roads in a certain geographic area.

portrayal
representation of geographic information to humans (ISO34 2002)
portrayal catalogue
collection of all defined portrayals (ISO49 2002)

portrayal rule
rule that is applied to the feature to determine what portrayal specification to use (ISO49 2002)

portrayal service
generic interface used to portray features (ISO49 2002)

portrayal specification
collection of operations applied to the feature instance to portray it (ISO49 2002)

positional accuracy
closeness of coordinate value to the true or accepted value in a specified reference system (ISO48 2002)
NOTE   The phrase "absolute accuracy" is sometimes used for this concept to distinguish it from relative positional accuracy. Where the true coordinate value may not be perfectly known, accuracy is normally tested by comparison to available values that can best be accepted as true.

positioning system
system of instrumental and computational components for determining position (ISO48 2002)
NOTE   Examples include inertial, integrated, linear, optical, and satellite positioning systems.

precision
measure of the repeatability of a set of measurements (ISO48 2002)
NOTE   Precision is usually expressed as a statistical value based upon a set of repeated measurements such as the standard deviation from the sample mean.

preferred term
term rated according to the scale of the term acceptability rating as the primary term for a given concept (ISO10 2000, ISO36 2002)

prime meridian (zero meridian)
meridian from which the longitudes of other meridians are quantified (ISO43 2003)
product specification
description of the universe of discourse and a specification for mapping the universe of discourse to a dataset (ISO45 2002)

profile
set of one or more base standards and — where applicable — the identification of chosen clauses, classes, options, and parameters of those base standards that are necessary for accomplishing a particular function (ISOIEC21 1998, ISO34 2002)
NOTE   A base standard is any ISO 19100 series standard or other Information Technology standard that can be used as a source for components from which a profile or product specification may be constructed

projected coordinate system
two-dimensional coordinate system resulting from a map projection (ISO43 2003)

quality

totality of characteristics of a product that bear on its ability to satisfy stated and implied needs (ISO34 2002)

quality schema
conceptual schema defining aspects of quality for geographic data (ISO34 2002)

record
finite, named collection of related items (objects or values) (ISO39 2003)
NOTE   Logically, a record is a set of pairs <name, item >.

reference data
data accepted as representing the universe of discourse, to be used as reference for direct external quality evaluation methods (ISO46 2001)

relative position
position of a point with respect to the positions of other points (ISO48 2002)
NOTE   The spatial relationship of one point relative to another may be single-, two- or three-dimensional.

relative positional accuracy
closeness of coordinate difference value to the true or accepted value in a specified reference system (ISO48 2002)
NOTE   Closely related terms such as local accuracy are employed in various countries, agencies, and application groups. Where such terms are utilized, a description of the term should be provided.

resource
asset or means that fulfils a requirement
EXAMPLE   Dataset, service, document, person or organization (ISO47 2003)

ring
simple curve which is a cycle
NOTE   Rings are used to describe boundary components of surfaces in 2D and 3D coordinate systems. (ISO39 2003)

satellite positioning system
positioning system based upon receipt of signals broadcast from satellites (ISO48 2002)
NOTE   In this context, satellite positioning implies the use of radio signals transmitted from "active" artificial objects orbiting the Earth and received by "passive" instruments on or near the Earth's surface to determine position, velocity, and/or attitude of an object. Examples are GPS and GLONASS.

schema
formal description of a model (ISO34 2002)

schema model
representation model for storing schemas (ISO50 2002)
EXAMPLE   Representation model for a schema repository

semi-major axis (*a*)
longest radius of an ellipsoid (ISO43 2003)
NOTE   For an ellipsoid representing the Earth, it is the radius of the equator.

semi-minor axis (*b*)
shortest radius of an ellipsoid (ISO43 2003)
NOTE   For an ellipsoid representing the Earth, it is the distance from the centre of the ellipsoid to either pole.

sequence
finite, ordered collection of related items (objects or values) that may be repeated (ISO39 2003)
NOTE   Logically, a sequence is a set of pairs <item, offset>. LISP syntax, which delimits sequences with parentheses and separates elements in the sequence with commas, is used in this International Standard.

service
capability which a service provider entity makes available to a service user entity at the interface between those entities (ISO34 2002)

service interface
shared boundary between an automated system or human being and another automated system or human being (ISO34 2002)

set
unordered collection of related items (objects or values) with no repetition (ISO39 2003)

shell
simple surface which is a cycle (ISO39 2003)
NOTE   Shells are used to describe boundary components of solids in 3D coordinate systems.

simple
property of a geometric object that its interior is isotropic (all points have isomorphic neighbourhoods), and hence everywhere locally isomorphic to an open subset of a Euclidean coordinate space of the appropriate dimension (ISO39 2003)
NOTE   This implies that no interior direct position is involved in a self-intersection of any kind.

simple feature

feature restricted to 2D geometry with linear interpolation between vertices, having both spatial and non spatial attributes (ISO54 2000)

solid
3-dimensional geometric primitive, representing the continuous image of a region of Euclidean 3 space (ISO39 2003)
NOTE   A solid is realizable locally as a 3 parameter set of direct positions. The boundary of a solid is the set of oriented, closed surfaces that comprise the limits of the solid.

spatial attribute
feature attribute describing the spatial representation of the feature by coordinates, mathematical functions and/or boundary topology relationships (ISO49 2002)

spatial object
instance of a type defined in the spatial schema (ISO34 2002)
object used for representing a spatial characteristic of a feature (ISO39 2003)

spatial operator
function or procedure that has at least one spatial parameter in its domain or range (ISO39 2003)
NOTE   Any UML operation on a spatial object would be classified as a spatial operator as are the query operators in clause 8 of ISO 19107.

spatial reference
description of position in the real world (ISO43 2003)
NOTE   This may take the form of a label, code or set of coordinates.

spatial reference system
system for identifying position in the real world (ISO44 2001)

start node
node in the boundary of an edge that corresponds to the start point of that edge as a curve in a valid geometric realization of the topological complex in which the edge is used (ISO39 2003)

start point
first point of a curve (ISO39 2003)

stereotype
new type of modelling element that extends the semantics of the metamodel (ISOIEC47 2003, ISO50 2002)
NOTE   Stereotypes must be based on certain existing types or classes in the metamodel. Stereotypes may extend the semantics, but not the structure of pre-existing types and classes. Certain stereotypes are perdefined in the UML, others

may be user defined. Stereotypes are one of three extensibility mechanisms in UML. The others are constraint and tagged value.

strong substitutability
ability for any instance of a class that is a descendant under inheritance or realization of another class, type or interface to be used in lieu of an instance of its ancestor in any context (ISO39 2003)
NOTE   The weaker forms of substitutability make various restrictions on the context of the implied substitution.

subcomplex
complex all of whose elements are also in a larger complex (ISO39 2003)
NOTE   Since the definitions of geometric complex and topological complex require only that they be closed under boundary operations, the set of any primitives of a particular dimension and below is always a subcomplex of the original, larger complex. Thus, any full planar topological complex contains an edge-node graph as a subcomplex.

surface
2-dimensional geometric primitive, locally representing a continuous image of a region of a plane (ISO39 2003)
NOTE   The boundary of a surface is the set of oriented, closed curves that delineate the limits of the surface. Surfaces that are isomorphic to a sphere, or to an n-torus (a topological sphere with n "handles") have no boundary. Such surfaces are called cycles.

surface patch
2-dimensional, connected geometric object used to represent a continuous portion of a surface using homogeneous interpolation and definition methods (ISO39 2003)

System Under Test (SUT)
computer hardware, software, and communication network required to support IUT (ISO37 2000)

temporal coordinate
distance from the origin of the interval scale used as the basis for a temporal coordinate system (ISO40 2002)

temporal coordinate system
temporal reference system based on an interval scale on which distance is measured as a multiple of a single unit of time (ISO40 2002)

temporal feature association
feature association characterized by a reference to time or to a temporal constraint (ISO40 2002)

Annex A: Terms and definitions of the ISO 19100 standards 257

temporal feature operation
feature operation specified as a function of time (ISO40 2002)

temporal position
location relative to a temporal reference system (ISO40 2002)

temporal reference system
reference system against which time is measured (ISO40 2002)

term
verbal designation of a general concept in specific subject field (ISO10 2000, ISO36 2002)
NOTE A term may contain symbols and can have variants, e.g. different forms of spelling.

term equivalent
term in another language which designates the same concept (ISO10 2000, ISO36 2002)

term instance classification
classification identifying the status of a term (ISO36 2002)

term repository
place where terms are stored or may be found (ISO36 2002)

terminological record
structured collection of terminological data relevant to one concept (ISO10 2000, ISO36 2002)

terminological record identifier
unique, unambiguous, and linguistically neutral identifier assigned to a term/definition (ISO36 2002)

testing laboratory
organization that carries out the conformance assessment process (ISO37 2000)

topological boundary
boundary represented by a set of oriented topological primitives of smaller topological dimension that limits the extent of a topological object (ISO39 2003)
NOTE The boundary of a topological complex corresponds to the boundary of the geometric realization of the topological complex.

topological complex
collection of topological primitives that is closed under the boundary operations (ISO39 2003)

NOTE	Closed under the boundary operations means that if a topological primitive is in the topological complex, then its boundary objects are also in the topological complex.

topological dimension
minimum number of free variables needed to distinguish nearby direct positions within a geometric object from one another (ISO39 2003)
NOTE The free variables mentioned above can usually be thought of as a local coordinate system. In a 3D coordinate space, a plane can be written as P(u, v) = A + u X + v Y, where u and v are real numbers and A is any point on the plane, and X and Y are two vectors tangent to the plane. since the locations on the plane can be distinguished by u and v (here universally), the plane is 2D and (u, v) is a coordinate system for the points on the plane. On generic surfaces, this cannot, in general, be done universally. If we take a plane tangent to the surface, and project points on the surface onto this plane, we will normally get a local isomorphism for small neighbourhoods of the point of tangency. This "local coordinate" system for the underlying surface is sufficient to establish the surface as a 2D topological object.
Since ISO 19107 deals only with spatial coordinates, any 3D object can rely on coordinates to establish its topological dimension. In a 4D model (spatio-temporal), tangent spaces also play an important role in establishing topological dimension for objects up to 3D.

topological expression
collection of oriented topological primitives which is operated upon like a multivariate polynomial (ISO39 2003)
NOTE	Topological expressions are used for many calculations in computational topology.

topological object
spatial object representing spatial characteristics that are invariant under continuous transformations (ISO39 2003)
NOTE	A topological object is a topological primitive, a collection of topological primitives, or a topological complex.

topological primitive
topological object that represents a single, non-decomposable element (ISO39 2003)
NOTE	A topological primitive corresponds to the interior of a geometric primitive of the same dimension in a geometric realization.

topological solid
3-dimensional topological primitive (ISO39 2003)
NOTE	The boundary of a topological solid consists of a set of directed faces.

transaction time
time when a fact is current in a database and may be retrieved (ISO40 2002)

transfer protocol
common set of rules for defining interactions between distributed systems (ISO50 2002)

uncertainty
parameter, associated with the result of measurement, that characterizes the dispersion of values that could reasonably be attributed to the measurand (ISO48 2002)
NOTE   When the quality of accuracy or precision of measured values, such as coordinates, is to be characterized quantitatively, the quality parameter is an estimate of the uncertainty of the measurement results. Because accuracy is a qualitative concept, one should not use it quantitatively, that is associate numbers with it; numbers should be associated with measures of uncertainty instead.

unit of measurement
defined quantity in which a dimensional parameter is expressed (ISO48 2002)
NOTE   In Positioning Services, the usual units of measurement are either angular units or linear units. Implementations of Positioning Services must clearly distinguish between SI units and non-SI units. When non-SI units are employed, their relation to SI units must be specified.

universal face
unbounded face in a 2-dimensional complex (ISO39 2003)
NOTE   The universal face is normally not part of any feature, and is used to represent the unbounded portion of the data set. Its interior boundary (it has no exterior boundary) would normally be considered the exterior boundary of the map represented by the data set. The ISO 19107 does not special case the universal face, but application schemas may find it convenient to do so.

universal solid
unbounded topological solid in a 3-dimensional complex (ISO39 2003)
NOTE   The universal solid is the 3-dimensional counterpart of the universal face, and is also normally not part of any feature.

universe of discourse
view of the real or hypothetical world that includes everything of interest (ISO34 2002)

valid time
time when a fact is true in the abstracted reality (ISO40 2002)

value domain
set of accepted values (ISO35 2001, ISO50 2002)
EXAMPLE   The range 3-28, all integers, any ASCII character, enumeration of all accepted values (green, blue, white).

vector geometry
representation of geometry through the use of constructive geometric primitives (ISO39 2003)

verification test
test developed to prove rigorously whether an IUT is correct (ISO37 2000)

vertical datum
datum describing the relation of gravity-related heights to the Earth (ISO43 2003)
NOTE   In most cases the vertical datum will be related to a defined mean sea level based on water level observations over a long time period. Ellipsoidal heights are treated as related to a three-dimensional ellipsoidal coordinate system referenced to a geodetic datum. Vertical datums include sounding datums (used for hydrographic purposes), in which case the heights may be negative heights or depths.

Annex B: ISO 19115 Metadata package data dictionaries

B.1 Metadata entity set information

	Name / Role name	Definition	Obligation / Condition
1.	MD_Metadata	root entity which defines metadata about a resource or resources	M
2.	fileIdentifier	unique identifier for this metadata file	O
3.	language	language used for documenting metadata	C / not defined by encoding?
4.	characterSet	full name of the character coding standard used for the metadata set	C / ISO/IEC 10646-1 not used and not defined by encoding
5.	parentIdentifier	file identifier of the metadata to which this metadata is a subset (child)	O
6.	hierarchyLevel	scope to which the metadata applies (see annex H for more information about metadata hierarchy levels)	C/ hierarchyLevel is not equal to "dataset"?
7.	hierarchyLevelName	name of the hierarchy levels for which the metadata is provided	O
8.	contact	party responsible for the metadata information	M
9.	dateStamp	date that the metadata was created	M
10.	metadataStandardName	name of the metadata standard (including profile name) used	O
11.	metadataStandardVersion	version (profile) of the metadata standard used	O
12.	Role name: spatialRepresentationInfo	digital representation of spatial information in the dataset	O
13.	Role name: referenceSystemInfo	description of the spatial and temporal reference systems used in the dataset	O

14.	Role name: metadataExtensionInfo	information describing metadata extensions	O
15.	Role name: identificationInfo	basic information about the resource(s) to which the metadata applies	M
16.	Role name: contentInfo	provides information about the feature catalogue and describes the coverage and image data characteristics	O
17.	Role name: distributionInfo	provides information about the distributor of and options for obtaining the resource(s)	O
18.	Role name: dataQualityInfo	provides overall assessment of quality of a resource(s)	O
19.	Role name: portrayalCatalogueInfo	provides information about the catalogue of rules defined for the portrayal of a resource(s)	O
20.	Role name: metadataConstraints	provides restrictions on the access and use of data	O
21.	Role name: applicationSchemaInfo	provides information about the conceptual schema of a dataset	O
22.	Role name: metadataMaintenance	provides information about the frequency of metadata updates, and the scope of those updates	O

B.2 Identification information (includes data and service identification)

	Name / Role name	Definition	Obligation / Condition
23.	MD_Identification	basic information required to uniquely identify a resource or resources	Use obligation from referencing object
24.	citation	citation data for the resource(s)	M
25.	abstract	brief narrative summary of the content of the resource(s)	M
26.	purpose	summary of the intentions with which the resource(s) was developed	O
27.	credit	recognition of those who contributed to the resource(s)	O
28.	status	status of the resource(s)	O
29.	pointOfContact	identification of, and means of communication with, person(s) and organizations(s) associated with the resource(s)	O
30.	Role name: resourceMaintenance	provides information about the frequency of resource updates, and the scope of those updates	O
31.	Role name: graphicOverview	provides a graphic that illustrates the resource(s) (should include a legend for the graphic)	O
32.	Role name: resourceFormat	provides a description of the format of the resource(s)	O
33.	Role name: descriptiveKeywords	provides category keywords, their type, and reference source	O
34.	Role name: resourceSpecificUsage	provides basic information about specific application(s) for which the resource(s) has/have been or is being used by different users	O

	Name / Role name	Definition	Obligation / Condition
35.	*Role name:* resourceConstraints	provides information about constraints which apply to the resource(s)	O
36.	MD_DataIdentification	information required to identify a dataset	Use obligation from referencing object
37.	spatialRepresentationType	method used to spatially represent geographic information	O
38.	spatialResolution	factor which provides a general understanding of the density of spatial data in the dataset	O
39.	language	language(s) used within the dataset	M
40.	characterSet	full name of the character coding standard used for the dataset	C/ISO/IEC 10646-1 not used?
41.	topicCategory	main theme(s) of the datset	M
42.	geographicBox	minimum bounding rectangle within which data is available	C / hierarchyLevel equals "dataset" and geographicDescription not documented?
43.	geographicDescription	description of the geographic area within which data is available	C / if hierarchyLevel equals "dataset" and geographicBox not documented?
44.	environmentDescription	description of the dataset in the producer's processing environment, including items such as the software, the computer operating system, file name, and the dataset size	O
45.	extent	additional extent information including the bounding polygon, vertical, and temporal extent of the dataset	O
46.	supplementalInformation	any other descriptive information about the dataset	O
47.	MD_ServiceIdentification	identification of capabilities which a service provider makes available to a service user through a set of interfaces that define a behaviour - See ISO 19119 – Services for further information	Use obligation from referencing object

B.2.1 Browse graphic information

	Name / Role name	Definition	Obligation / Condition
48.	MD_BrowseGraphic	graphic that provides an illustration of the dataset (should include a legend for the graphic)	Use obligation from referencing object
49.	fileName	name of the file that contains a graphic that provides an illustration of the dataset	M
50.	fileDescription	text description of the illustration	O

Annex B: ISO 19115 Metadata package data dictionaries 263

51.	fileType	format in which the illustration is encoded Examples: CGM, EPS, GIF, JPEG, PBM, PS, TIFF, XWD	O

B.2.2 Keyword information

52.	MD_Keywords	keywords, their type and reference source	Use obligation from referencing object
53.	keyword	commonly used word(s) or formalised word(s) or phrase(s) used to describe the subject	M
54.	type	subject matter used to group similar keywords	O
55.	thesaurusName	name of the formally registered thesaurus or a similar authoritative source of keywords	O

B.2.3 Representative fraction information

56.	MD_RepresentativeFraction	derived from Scale where MD_RepresentativeFraction.denominat or = 1 / Scale.measure And Scale.targetUnits = Scale.sourceUnits	Use obligation from referencing object
57.	denominator	the number below the line in a vulgar fraction	M
58.	Role name (derived): /Scale	role indicating that MD_RepresentativeFraction is derived from Scale	(Not applicable)

B.2.4 Resolution information

59.	MD_Resolution	level of detail expressed as a scale factor or a ground distance	Use obligation from referencing object
60.	equivalentScale	level of detail expressed as the scale of a comparable hardcopy map or chart	C / distance not documented?
61.	distance	ground sample distance	C / equivalentScale not documented?

B.2.5 Usage information

62.	MD_Usage	brief description of ways in which the resource(s) is/are currently used	Use obligation from referencing object
63.	specificUsage	brief description of the resource and/or resource series usage	M
64.	usageDateTime	date and time of the first use or range of uses of the resource and/or resource series	O

	Name	Definition	Obligation / Condition
65.	userDeterminedLimitations	applications, determined by the user for which the resource and/or resource series is not suitable	O
66.	userContactInfo	identification of and means of communicating with person(s) and organization(s) using the resource(s)	M

B.3 Constraint information (includes legal and security)

	Name	Definition	Obligation / Condition
67.	MD_Constraints	restrictions on the access and use of a resource or metadata	Use obligation from referencing object
68.	useLimitation	limitation affecting the fitness for use of the resource. Example, "not to be used for navigation"	O
69.	MD_LegalConstraints	restrictions and legal prerequisites for accessing and using the resource	Use obligation from referencing object
70.	accessConstraints	access constraints applied to assure the protection of privacy or intellectual property, and any special restrictions or limitations on obtaining the resource	O
71.	useConstraints	constraints applied to assure the protection of privacy or intellectual property, and any special restrictions or limitations or warnings on using the resource	O
72.	otherConstraints	other restrictions and legal prerequisites for accessing and using the resource	C / accessConstraints or useConstraints equal "otherRestrictions"?
73.	MD_SecurityConstraints	handling restrictions imposed on the resource for national security or similar security concerns	Use obligation from referencing object
74.	classification	name of the handling restrictions on the resource	M
75.	userNote	explanation of the application of the legal constraints or other restrictions and legal prerequisites for obtaining and using the resource	O
76.	classificationSystem	name of the classification system	O
77.	handlingDescription	additional information about the restrictions on handling the resource	O

B.4 Data quality information

	Name	Definition	Obligation / Condition
78.	DQ_DataQuality	quality information for the data specified by a data quality scope	Use obligation from referencing object

Annex B: ISO 19115 Metadata package data dictionaries

79.	scope	the specific data to which the data quality information applies	M
80.	*Role name:* report	quantitative quality information for the data specified by the scope	C / scope.DQ_Scope. level equals "dataset"?
81.	*Role name:* lineage	non-quantitative quality information about the lineage of the data specified by the scope	C / scope.DQ_Scope. level equals "dataset"?

B.4.1 Lineage information

82.	LI_Lineage	information about the events or source data used in constructing the data specified by the scope or lack of knowledge about lineage	Use obligation from referencing object
83.	statement	general explanation of the data producer's knowledge about the lineage of a dataset	C / (DQ_DataQuality. scope.DQ_Scope. level = "dataset" or "series") and source and processStep not provided?
84.	*Role name:* processStep	information about an event in the creation process for the data specified by the scope	C / statement and source not provided?
85.	*Role name:* source	information about the source data used in creating the data specified by the scope	C / statement and processStep not provided?

B.4.1.1 Process step information

86.	LI_ProcessStep	information about an event in the creation process for the data specified by the scope	Use obligation from referencing object
87.	description	description of the event, including related parameters or tolerances	M
88.	rationale	requirement or purpose for the process step	O
89.	dateTime	date and time or range of date and time on or over which the process step occurred	O
90.	processor	identification of, and means of communication with, person(s) and organization(s) associated with the process step	O
91.	*Role name:* source	information about the source data used in creating the data specified by the scope	O

B.4.1.2 Source information

92.	LI_Source	information about the source data used in creating the data specified by the scope	Use obligation from referencing object
93.	description	detailed description of the level of the source data	C / sourceExtent not provided?
94.	scaleDenominator	denominator of the representative fraction on a source map	O
95.	sourceReferenceSystem	spatial reference system used by the source data	O
96.	sourceCitation	recommended reference to be used for the source data	O
97.	sourceExtent	information about the spatial, vertical and temporal extent of the source data	C / description not provided?
98.	Role name: sourceStep	information about an event in the creation process for the source data	O

B.4.2 Data quality element information

99.	DQ_Element	type of test applied to the data specified by a data quality scope	Use obligation from referencing object
100.	nameOfMeasure	name of the test applied to the data	O
101.	measureIdentification	code identifying a registered standard procedure	O
102.	measureDescription	description of the measure being determined	O
103.	evaluationMethodType	type of method used to evaluate quality of the dataset	O
104.	evaluationMethodDescription	description of the evaluation method	O
105.	evaluationProcedure	reference to the procedure information	O
106.	dateTime	date or range of dates on which a data quality measure was applied	O
107.	result	value (or set of values) obtained from applying a data quality measure or the out come of evaluating the obtained value (or set of values) against a specified acceptable conformance quality level	M
108.	DQ_Completeness	presence and absence of features, their attributes and their relationships	Use obligation from referencing object
109.	DQ_CompletenessCommission	excess data present in the dataset, as described by the scope	Use obligation from referencing object
110.	DQ_CompletenessOmission	data absent from the dataset, as described by the scope	Use obligation from referencing object

… Annex B: ISO 19115 Metadata package data dictionaries

111.	DQ_LogicalConsistency	degree of adherence to logical rules of data structure, attribution and relationships (data structure can be conceptual, logical or physical)	Use obligation from referencing object
112.	DQ_ConceptualConsistency	adherence to rules of the conceptual schema	Use obligation from referencing object
113.	DQ_DomainConsistency	adherence of values to the value domains	Use obligation from referencing object
114.	DQ_FormatConsistency	degree to which data is stored in accordance with the physical structure of the dataset, as described by the scope	Use obligation from referencing object
115.	DQ_TopologicalConsistency	correctness of the explicitly encoded topological characteristics of the dataset as described by the scope	Use obligation from referencing object
116.	DQ_PositionalAccuracy	accuracy of the position of features	Use obligation from referencing object
117.	DQ_AbsoluteExternalPositionalAccuracy	closeness of reported coordinate values to values accepted as or being true	Use obligation from referencing object
118.	DQ_GriddedDataPositionalAccuracy	closeness of gridded data position values to values accepted as or being true	Use obligation from referencing object
119.	DQ_RelativeInternalPositionalAccuracy	closeness of the relative positions of features in the scope to their respective relative positions accepted as or being true	Use obligation from referencing object
120.	DQ_TemporalAccuracy	accuracy of the temporal attributes and temporal relationships of features	Use obligation from referencing object
121.	DQ_AccuracyOfATimeMeasurement	correctness of the temporal references of an item (reporting of error in time measurement)	Use obligation from referencing object
122.	DQ_TemporalConsistency	correctness of ordered events or sequences, if reported	Use obligation from referencing object
123.	DQ_TemporalValidity	validity of data specified by the scope with respect to time	Use obligation from referencing object
124.	DQ_ThematicAccuracy	accuracy of quantitative attributes and the correctness of non-quantitative attributes and of the classifications of features and their relationships	Use obligation from referencing object
125.	DQ_ThematicClassificationCorrectness	comparison of the classes assigned to features or their attributes to a universe of discourse	Use obligation from referencing object
126.	DQ_NonQuantitativeAttributeAccuracy	correctness of non-quantitative attributes	Use obligation from referencing object
127.	DQ_QuantitativeAttributeAccuracy	accuracy of quantitative attributes	Use obligation from referencing object

B.4.3 Result information

128.	DQ_Result	generalization of more specific result classes	Use obligation from referencing object
129.	DQ_ConformanceResult	Information about the outcome of evaluating the obtained value (or set of values) against a specified acceptable conformance quality level	Use obligation from referencing object
130.	specification	citation of product specification or user requirement against which data is being evaluated	M
131.	explanation	explanation of the meaning of conformance for this result	M
132.	pass	indication of the conformance result where 0 = fail and 1 = pass	M
133.	DQ_QuantitativeResult	Information about the value (or set of values) obtained from applying a data quality measure	Use obligation from referencing object
134.	valueType	value type for reporting a data quality result	O
135.	valueUnit	value unit for reporting a data quality result	O
136.	errorStatistic	statistical method used to determine the value	O
137.	value	quantitative value or values, content determined by the evaluation procedure used	M

B.4.4 Scope information

138.	DQ_Scope	description of the data specified by the scope	Use obligation from referencing object
139.	level	hierarchical level of the data specified by the scope	M
140.	extent	information about the spatial, vertical and temporal extent of the data specified by the scope	O
141.	levelDescription	detailed description about the level of the data specified by the scope	C / level not equal "dataset" or "series"?

B.5 Maintenance information

	Name	Definition	Obligation / Condition
142.	MD_MaintenanceInformation	information about the scope and frequency of updating	Use obligation from referencing object
143.	maintenanceAndUpdateFrequency	frequency with which changes and additions are made to the resource after the initial resource is completed	M

144.	dateOfNextUpdate	scheduled revision date for resource	O
145.	userDefinedMaintenanceFrequency	maintenance period other than those defined	O
146.	updateScope	scope of data to which maintenance is applied	O
147.	updateScopeDescription	additional information about the range or extent of the resource	O
148.	maintenanceNote	information regarding specific requirements for maintaining the resource	O

B.5.1 Scope description information

149.	MD_ScopeDescription	description of the class of information covered by the information	Use obligation from referencing object
150.	attributes	attributes to which the information applies	C / features, featureInstances, attributeInstances, dataset and other not documented?
151.	features	features to which the information applies	C / attributes, featureInstances, attributeInstances, dataset and other not documented?
152.	featureInstances	feature instances to which the information applies	C / attributes, features, attributeInstances, dataset and other not documented?
153.	attributeInstances	attribute instances to which the information applies	C / attributes, features, featureInstances, dataset and other not documented?
154.	dataset	dataset to which the information applies	C / attributes, features, featureInstances, attributeInstances, and other not documented?
155.	other	class of information that does not fall into the other categories to which the information applies	C / attributes, features, featureInstances, attributeInstances, and dataset not documented?

B.6 Spatial representation information (includes grid and vector representation)

	Name	Definition	Obligation / Condition

156.	MD_SpatialRepresentation	digital mechanism used to represent spatial information	Use obligation/condition from referencing object
157.	MD_GridSpatialRepresentation	information about grid spatial objects in the dataset	Use obligation/condition from referencing object
158.	numberOfDimensions	number of independent spatial-temporal axes	M
159.	axisDimensionsProperties	information about spatial-temporal axis properties	M
160.	cellGeometry	identification of grid data as point or cell	M
161.	transformationParameterAvailability	indication of whether or not parameters for transformation exists	M
162.	MD_Georectified	grid whose cells are regularly spaced in a geographic (i.e., lat / long) or map coordinate system defined in the Spatial Referencing System (SRS) so that any cell in the grid can be geolocated given its grid coordinate and the grid origin, cell spacing, and orientation	Use obligation/condition from referencing object
163.	checkPointAvailability	indication of whether or not geographic position points are available to test the accuracy of the georeferenced grid data	M
164.	checkPointDescription	description of geographic position points used to test the accuracy of the georeferenced grid data	C / checkPointAvailability equals "yes"?
165.	cornerPoints	earth location in the coordinate system defined by the Spatial Reference System and the grid coordinate of the cells at opposite ends of grid coverage along two diagonals in the grid spatial dimensions. There are four corner points in a georectified grid; at least two corner points along one diagonal are required	M
166.	centerPoint	earth location in the coordinate system defined by the Spatial Reference System and the grid coordinate of the cell halfway between opposite ends of the grid in the spatial dimensions	O
167.	pointInPixel	point in a pixel corresponding to the Earth location of the pixel	M
168.	transformationDimensionDescription	description of the information about which grid dimensions are the spatial dimensions	O
169.	transformationDimensionMapping	information about which grid dimensions are the spatial dimensions	O
170.	MD_Georeferenceable	grid with cells irregularly spaced in any given geographic/map projection coordinate system, whose individual cells can be geolocated using geolocation information supplied with the data but cannot be geolocated from the grid properties alone	Use obligation/condition from referencing object

Annex B: ISO 19115 Metadata package data dictionaries   271

171.	controlPointAvailability	indication of whether or not control point(s) exists	M
172.	orientationParameterAvailability	indication of whether or not orientation parameters are available	M
173.	orientationParameterDescription	description of parameters used to describe sensor orientation	O
174.	parameters	terms which support grid data georeferencing	M
175.	parameterCitation	reference providing description of the parameters	O
176.	MD_VectorSpatialRepresentation	information about the vector spatial objects in the dataset	Use obligation/condition from referencing object
177.	topologyLevel	code which identifies the degree of complexity of the spatial relationships	O
178.	geometricObjects	information about the geometric objects used in the dataset	O

B.6.1 Dimension information

179.	MD_Dimension	axis properties	Use obligation/condition from referencing object
180.	dimensionName	name of the axis	M
181.	dimensionSize	number of elements along the axis	M
182.	resolution	degree of detail in the grid dataset	O

B.6.2 Geometric object information

183.	MD_GeometricObjects	number of objects, listed by geometric object type, used in the dataset	Use obligation/condition from referencing object
184.	geometricObjectType	name of point and vector spatial objects used to locate zero-, one-, and two-dimensional spatial locations in the dataset	M
185.	geometricObjectCount	total number of the point or vector object type occurring in the dataset	O

B.7 Reference system information (includes temporal, coordinate and geographic identifiers)

	Name	Definition	Obligation / Condition
186.	MD_ReferenceSystem	information about the reference system.	Use obligation/condition from referencing object

272   Annexes

187.	referenceSystemIdentifier	name of reference system	C / MD_CRS.projection, MD_CRS.ellipsoid, and MD_CRS.datum not documented?
188.	role name (derived): /Reference System	relationship indicating that MD_ReferenceSystem (as well as its attributes and aggregates) is derived from RS_ReferenceSystem	(Not applicable)
189.	MD_CRS	metadata about a coordinate system in which attributes have been derived from SC_CRS as defined in ISO 19111 – Spatial referencing by coordinates	Use obligation/condition from referencing object
190.	projection	identity of the projection used	O
191.	ellipsoid	identity of the ellipsoid used	O
192.	datum	Identity of the datum used	O
193.	role name: ellipsoidParameters	set of parameters that describe the ellipsoid	O
194.	role name: projectionParameters	set of parameters that describe the projection	O
195.	RS_ReferenceSystem	description of the spatial and temporal reference systems used in the dataset	Use obligation/condition from referencing object
196.	name	name of reference system used	M
197.	domainOfValidity	range which is valid for the reference system	O
198.	TM_ReferenceSystem	documented in ISO 19108 – Temporal schema	Use obligation/condition from referencing object
199.	SI_SpatialReferenceSystemUsingGeographicIdentifiers	documented in ISO 19112 – Spatial referencing by geographic identifiers	Use obligation/condition from referencing object
200.	SC_CRS	documented in ISO 19111 – Spatial referencing by coordinates	Use obligation/condition from referencing object

B.7.1 Ellipsoid parameter information

201.	MD_EllipsoidParameters	set of parameters that describe the ellipsoid	Use obligation/condition from referencing object
202.	semiMajorAxis	radius of the equatorial axis of the ellipsoid	M
203.	axisUnits	units of the semi-major axis	M

Annex B: ISO 19115 Metadata package data dictionaries 273

204.	denominatorOfFlatteningRatio	ratio of the difference between the equatorial and polar radii of the ellipsoid to the equatorial radius when the numerator is set to 1	C / not a spheroid?

B.7.2 Identifier information

205.	MD_Identifier	value uniquely identifying an object within a namespace	Use obligation/condition from referencing object
206.	authority	person or party responsible for maintenance of the namespace	O
207.	code	alphanumeric value identifying an instance in the namespace	M
208.	RS_Identifier	identifier used for reference systems	Use obligation/condition from referencing object

B.7.3 Oblique line azimuth information

209.	MD_ObliqueLineAzimuth	method used to describe the line along which an oblique mercator map projection is centred using the map projection origin and an azimuth	Use obligation/condition from referencing object
210.	azimuthAngle	angle measured clockwise from north, and expressed in degrees	M
211.	azimuthMeasurePointLongitude	longitude of the map projection origin	M

B.7.4 Oblique line point information

212.	MD_ObliqueLinePoint	method used to describe the line along which an oblique mercator map projection is centred using two points near the limits of the mapped region that define the centre line	Use obligation/condition from referencing object
213.	obliqueLineLatitude	latitude of a point defining the oblique line	M
214.	obliqueLineLongitude	longitude of a point defining the oblique line	M

B.7.5 Projection parameter information

215.	MD_ProjectionParameters	set of parameters that describe the projection	Use obligation/condition from referencing object
216.	zone	unique identifier for 100,000 metre grid zone	O

274 Annexes

217.	standardParallel	line of constant latitude at which the surface of the Earth and the plane or developable surface intersect	O
218.	longitudeOfCentralMeridian	line of longitude at the centre of a map projection generally used as the basis for constructing the projection	O
219.	latitudeOfProjectionOrigin	latitude chosen as the origin of rectangular coordinates for a map projection	O
220.	falseEasting	value added to all "x" values in the rectangular coordinates for a map projection. This value frequently is assigned to eliminate negative numbers. Expressed in the unit of measure identified in Planar Coordinate Units	O
221.	falseNorthing	value added to all "y" values in the rectangular coordinates for a map projection. This value frequently is assigned to eliminate negative numbers. Expressed in the unit of measure identified in Planar Coordinate Units	O
222.	falseEastingNorthingUnits	units of false northing and false easting	O
223.	scaleFactorAtEquator	ratio between physical distance and corresponding map distance, along the equator	O
224.	heightOfProspectivePointAboveSurface	height of viewpoint above the Earth, expressed in metres	O
225.	longitudeOfProjectionCenter	longitude of the point of projection for azimuthal projections	O
226.	latitudeOfProjectionCenter	latitude of the point of projection for azimuthal projections	O
227.	scaleFactorAtCenterLine	ratio between physical distance and corresponding map distance, along the centre line	O
228.	straightVerticalLongitudeFromPole	longitude to be oriented straight up from the North or South Pole	O
229.	scaleFactorAtProjectionOrigin	multiplier for reducing a distance obtained from a map by computation or scaling tot he actual distance at the projection origin	O
230.	*role name:* obliqueLineAzimuthParameter	parameters describing the oblique line azimuth	O
231.	*role name:* obliqueLinePointParameter	parameters describing the oblique line point	O

B.8 Content information (includes Feature catalogue and Coverage descriptions)

	Name / Role name	Definition	Obligation / Condition

Annex B: ISO 19115 Metadata package data dictionaries 275

232.	MD_ContentInformation	description of the content of a dataset	Use obligation/condition from referencing object
233.	MD_FeatureCatalogueDescription	information identifying the feature catalogue	Use obligation/condition from referencing object
234.	complianceCode	indication of whether or not the cited feature catalogue complies with ISO 19110	O
235.	language	language(s) used within the catalogue	O
236.	includedWithDataset	indication of whether or not the feature catalogue is included with the dataset	M
237.	featureTypes	subset of feature types from cited feature catalogue occurring in dataset	O
238.	featureCatalogueCitation	complete bibliographic reference to one or more external feature catalogues	M
239.	MD_CoverageDescription	information about the content of a grid data cell	Use obligation/condition from referencing object
240.	attributeDescription	description of the attribute described by the measurement value	M
241.	contentType	type of information represented by the cell value	M
242.	*Role name*: dimension	information on the dimensions of the cell measurement value	O
243.	MD_ImageDescription	information about an image's suitability for use	O
244.	illuminationElevationAngle	illumination elevation measured in degrees clockwise from the target plane at intersection of the optical line of sight with the Earth's surface. For images from a scanning device, refer to the centre pixel of the image	O
245.	illuminationAzimuthAngle	illumination azimuth measured in degrees clockwise from true north at the time the image is taken. For images from a scanning device, refer to the centre pixel of the image	O
246.	imagingCondition	conditions affected the image	O
247.	imageQualityCode	specifies the image quality	O
248.	cloudCoverPercentage	area of the dataset obscured by clouds, expressed as a percentage of the spatial extent	O
249.	processingLevelCode	image distributor's code that identifies the level of radiometric and geometric processing that has been applied	O
250.	compressionGenerationQuantity	count of the number the number of lossy compression cycles performed on the image	O
251.	triangulationIndicator	indication of whether or not triangulation has been performed upon the image	O

252.	radiometricCalibration-DataAvailability	indication of whether or not the radiometric calibration information for generating the radiometrically calibrated standard data product is available	O
253.	cameraCalibrationInformationAvailability	indication of whether or not constants are available which allow for camera calibration corrections	O
254.	filmDistortionInformationAvailability	indication of whether or not Calibration Reseau information is available	O
255.	lensDistortionInformationAvailability	indication of whether or not lens aberration correction information is available	O

B.8.1 Range dimension information (includes Band information)

256.	MD_RangeDimension	information on the range of each dimension of a cell measurement value	Use obligation/condition from referencing object
257.	sequenceIdentifier	number that uniquely identifies instances of bands of wavelengths on which a sensor operates	O
258.	descriptor	description of the range of a cell measurement value	O
259.	MD_Band	range of wavelengths in the electromagnetic spectrum	Use obligation/condition from referencing object
260.	maxValue	longest wavelength that the sensor is capable of collecting within a designated band	O
261.	minValue	shortest wavelength that the sensor is capable of collecting within a designated band	O
262.	units	units in which sensor wavelengths are expressed	C / minValue or maxValue provided?
263.	peakResponse	wavelength at which the response is the highest	O
264.	bitsPerValue	maximum number of significant bits in the uncompressed representation for the value in each band of each pixel	O
265.	toneGradation	number of discrete numerical values in the grid data	O
266.	scaleFactor	scale factor which has been applied to the cell value	O
267.	offset	the physical value corresponding to a cell value of zero	O

B.9 Portrayal catalogue information

	Name	Definition	Obligation / Condition

268.	MD_PortrayalCatalogueReference	information identifying the portrayal catalogue used	Use obligation/condition from referencing object
269.	portrayalCatalogueCitation	bibliographic reference to the portrayal catalogue cited	M

B.10 Distribution information

	Name / Role name	Definition	Obligation / Condition
270.	MD_Distribution	information about the distributor of and options for obtaining the resource	Use obligation/condition from referencing object
271.	*Role name:* distributionFormat	provides a description of the format of the data to be distributed	C / MD_Distributor.distibutorFormat not documented?
272.	*Role name:* distributor	provides information about the distributor	O
273.	*Role name:* transferOptions	provides information about technical means and media by which a resource is obtained from the distributor	O

B.10.1 Digital transfer options information

274.	MD_DigitalTransferOptions	technical means and media by which a resource is obtained from the distributor	Use obligation/condition from referencing object
275.	unitsOfDistribution	tiles, layers, geographic areas, etc., in which data is available	O
276.	transferSize	estimated size of a unit in the specified transfer format, expressed in megabytes. The transfer size is > 0.0	O
277.	onLine	information about online sources from which the resource can be obtained	O
278.	offLine	information about offline media on which the resource can be obtained	O

B.10.2 Distributor information

279.	MD_Distributor	information about the distributor	Use obligation/condition from referencing object
280.	distributorContact	party from whom the resource may be obtained. This list need not be exhaustive	M
281.	*Role name:* distributionOrderProcess	provides information about how the resource may be obtained, and related instructions and fee information	O

282.	Role name: distributorFormat	provides information about the format used by the distributor	C / MD_Distribution.distributionFormat not documented?
283.	Role name: distributorTransferOptions	provides information about the technical means and media used by the distributor	O

B.10.3 Format information

284.	MD_Format	description of the computer language construct that specifies the representation of data objects in a record, file, message, storage device or transmission channel	Use obligation/condition from referencing object
285.	name	name of the data transfer format(s)	M
286.	version	version of the format (date, number, etc.)	M
287.	amendmentNumber	amendment number of the format version	O
288.	specification	name of a subset, profile, or product specification of the format	O
289.	fileDecompressionTechnique	recommendations of algorithms or processes that can be applied to read or expand resources to which compression techniques have been applied	O
290.	Role name: formatDistributor	provides information about the distributor's format	O

B.10.4 Medium information

291.	MD_Medium	information about the media on which the resource can be distributed	Use obligation/condition from referencing object
292.	name	name of the medium on which the resource can be received	O
293.	density	density at which the data is recorded	O
294.	densityUnits	units of measure for the recording density	C / density documented?
295.	volumes	number of items in the media identified	O
296.	mediumFormat	method used to write to the medium	O
297.	mediumNote	description of other limitations or requirements for using the medium	O

B.10.5 Standard order process information

	MD_StandardOrderProcess	common ways in which the resource may be obtained or received, and related instructions and fee information	Use obligation/condition from referencing object

298.	fees	fees and terms for retrieving the resource. Include monetary units (as specified in ISO 4217)	O
299.	plannedAvailableDateTime	date and time when the dataset will be available	O
300.	orderingInstructions	general instructions, terms and services provided by the distributor	O
301.	turnaround	typical turnaround time for the filling of an order	O

B.11 Metadata extension information

	Name	Definition	Obligation / Condition
302.	MD_MetadataExtensionInformation	information describing metadata extensions	Use obligation/condition from referencing object
303.	extensionOnLineResource	information about on-line sources containing the community profile name and the extended metadata elements. Information for all new metadata elements	O
304.	*Role name:* extendedElementInformation	provides information about a new metadata element, not found in ISO 19115, which is required to describe geographic data	O

B.11.1 Extended element information

305.	MD_ExtendedElementInformation	new metadata element, not found in ISO 19115, which is required to describe geographic data	Use obligation/condition from referencing object
306.	name	name of the extended metadata element.	M
307.	shortName	short form suitable for use in an implementation method such as XML or SGML. NOTE other methods may be used	C / dataType notEqual "codelistElement"?
308.	domainCode	three digit code assigned to the extended element	C / is dataType "codelistElement"?
309.	definition	definition of the extended element	M
310.	obligation	obligation of the extended element	C / dataType not "codelist", "enumeration" or "codelistElement"?
311.	condition	condition under which the extended element is mandatory	C / obligation = "Conditional"?
312.	dataType	code which identifies the kind of value provided in the extended element	M

313.	maximumOccurrence	maximum occurrence of the extended element	C / dataType not "codelist", "enumeration" or "codelistElement"?
314.	domainValue	valid values that can be assigned to the extended element	C / dataType not "codelist", "enumeration" or "codelistElement"?
315.	parentEntity	name of the metadata entity(s) under which this extended metadata element may appear. The name(s) may be standard metadata element(s) or other extended metadata element(s).	M
316.	rule	specifies how the extended element relates to other existing elements and entities	M
317.	rationale	reason for creating the extended element	O
318.	source	name of the person or organization creating the extended element	M

B.12 Application schema information

	Name	Definition	Obligation / Condition
319.	MD_ApplicationSchemaInformation	information about the application schema used to build the dataset	Use obligation/condition from referencing object
320.	name	name of the application schema used	M
321.	schemaLanguage	identification of the schema language used	M
322.	constraintLanguage	formal language used in Application Schema	M
323.	schemaAscii	full application schema given as an ASCII file	O
324.	graphicsFile	full application schema given as a graphics file	O
325.	softwareDevelopmentFile	full application schema given as a software development file	O
326.	softwareDevelopmentFile-Format	software dependent format used for the application schema software dependent file	O
327.	*Role name:* featureCatalogueSupplement	information about the spatial attributes in the application schema for the feature types	O

Annex B: ISO 19115 Metadata package data dictionaries 281

B.12.1 Feature type list information

328.	MD_FeatureTypeList	list of names of feature types with the same spatial representation (same as spatial attributes)	Use obligation/condition from referencing object
329.	spatialObject	instance of a type defined in the spatial schema	M
330.	spatialSchemaName	name of the spatial schema used	M

B.12.2 Spatial attribute supplement information

331.	MD_SpatialAttributeSupplement	spatial attributes in the application schema for the feature types	Use obligation/condition from referencing object
332.	*Role name:* theFeatureTypeList	provides information about the list of feature types with the same spatial representation	M

B.13 Extent information

	Name / Role name	Definition	Obligation / Condition
333.	EX_Extent	information about spatial, vertical, and temporal extent	Use obligation/condition from referencing object
334.	description	spatial and temporal extent for the referring object	C / geographicElement and temporalElement and verticalElement not documented?
335.	*Role name:* geographicElement	provides geographic component of the extent of the referring object	C / description and temporalElement and verticalElement not documented?
336.	*Role name:* temporalElement	provides temporal component of the extent of the referring object	C / description and geographicElement and verticalElement not documented?
337.	*Role name:* verticalElement	provides vertical component of the extent of the referring object	C / description and geographicElement and temporalElement not documented?

B.13.1 Geographic extent information

338.	EX_GeographicExtent	geographic area of the dataset	Use obligation/condition from referencing object
339.	extentTypeCode	indication of whether the bounding polygon encompasses an area covered by the data or an area where data is not present	O
340.	EX_BoundingPolygon	boundary enclosing the dataset, expressed as the closed set of (x,y) coordinates of the polygon (last point replicates first point)	Use obligation/condition from referencing object
341.	polygon	sets of points defining the bounding polygon	M
342.	EX_GeographicBoundingBox	geographic position of the dataset NOTE This is only an approximate reference so specifying the co-ordinate system is unnecessary	Use obligation/condition from referencing object
343.	westBoundLongitude	western-most coordinate of the limit of the dataset extent, expressed in longitude in decimal degrees (positive east)	M
344.	eastBoundLongitude	eastern-most coordinate of the limit of the dataset extent, expressed in longitude in decimal degrees (positive east)	M
345.	southBoundLatitude	southern-most coordinate of the limit of the dataset extent, expressed in latitude in decimal degrees (positive north)	M
346.	northBoundLatitude	northern-most, coordinate of the limit of the dataset extent expressed in latitude in decimal degrees (positive north)	M
347.	EX_GeographicDescription	Description of the geographic area using identifiers	Use obligation/condition from referencing object
348.	geographicIdentifier	identifier used to represent a geographic area	M

B.13.2 Temporal extent information

349.	EX_TemporalExtent	time period covered by the content of the dataset	Use obligation/condition from referencing object
350.	extent	date and time for the content of the dataset	M
351.	EX_SpatialTemporalExtent	extent with respect to date/time and spatial boundaries	Use obligation/condition from referencing object
352.	role name: spatialExtent	spatial extent component of composite spatial and temporal extent	M

B.13.3 Vertical extent information

353.	EX_VerticalExtent	vertical domain of dataset	Use obligation/condition from referencing object
354.	minimumValue	lowest vertical extent contained in the dataset	M
355.	maximumValue	highest vertical extent contained in the dataset	M
356.	unitOfMeasure	vertical units used for vertical extent information Examples: metres, feet, millimetres, hectopascals	M
357.	*role name:* verticalDatum	provides information about the origin from which the maximum and minimum elevation values are measured	M

B.14 Citation and responsible party information

	Name / Role name	Definition	Obligation / Condition
358.	CI_Citation	standardized resource reference	Use obligation/condition from referencing object
359.	title	name by which the cited resource is known	M
360.	alternateTitle	short name or other language name by which the cited information is known. Example: "DCW" as an alternative title for "Digital Chart of the World"	O
361.	date	reference date for the cited resource	M
362.	edition	version of the cited resource	O
363.	editionDate	date of the edition	O
364.	identifier	unique identifier for the resource EXAMPLE: Universal Product Code (UPC), National Stock Number (NSN)	O
365.	identifierType	reference form of the unique identifier (ID) Example: Universal Product Code (UPC), National Stock Number (NSN)	O
366.	citedResponsibleParty	name and position information for an individual or organization that is responsible for the resource	O
367.	presentationForm	mode in which the resource is represented	O
368.	series	information about the series, or aggregate dataset, of which the dataset is a part	O
369.	otherCitationDetails	other information required to complete the citation that is not recorded elsewhere	O

370.	collectiveTitle	common title with holdings note NOTE title identifies elements of a series collectively, combined with information about what volumes are available at the source cited	O
371.	ISBN	international Standard Book Number	O
372.	ISSN	international Standard Serial Number	O
373.	CI_ResponsibleParty	identification of, and means of communication with, person(s) and organizations associated with the dataset	Use obligation/condition from referencing object
374.	individualName	name of the responsible person- surname, given name, title separated by a delimiter	C / organisationName and positionName not documented?
375.	organisationName	name of the responsible organization	C / individualName and positionName not documented?
376.	positionName	role or position of the responsible person	C / individualName and organisationName not documented?
377.	contactInfo	address of the responsible party	O
378.	role	function performed by the responsible party	M

B.14.1 Address information

379.	CI_Address	location of the responsible individual or organization	Use obligation/condition from referencing object
380.	deliveryPoint	address line for the location (as described in ISO 11180, annex A)	O
381.	city	city of the location	O
382.	administrativeArea	state, province of the location	O
383.	postalCode	ZIP or other postal code	O
384.	country	country of the physical address	O
385.	electronicMailAddress	address of the electronic mailbox of the responsible organization or individual	O

B.14.2 Contact information

386.	CI_Contact	information required to enable contact with the responsible person and/or organization	Use obligation/condition from referencing object
387.	phone	telephone numbers at which the organization or individual may be contacted	O
388.	address	physical and email address at which the organization or individual may be contacted	O
389.	onLineResource	on-line information that can be used to contact the individual or organization	O

| 390. | hoursOfService | time period (including time zone) when individuals can contact the organization or individual | O |
| 391. | contactInstructions | supplemental instructions on how or when to contact the individual or organization | O |

B.14.3 Date information

392.	CI_Date	reference date and event used to describe it	Use obligation/condition from referencing object
393.	date	reference date for the cited resource	M
394.	dateType	event used for reference date	M

B.14.4 OnLine resource information

395.	CI_OnLineResource	information about on-line sources from which the dataset, specification, or community profile name and extended metadata elements can be obtained	Use obligation/condition from referencing object
396.	linkage	location (address) for on-line access using a Uniform Resource Locator address or similar addressing scheme such as http://www.statkart.no/isotc211	M
397.	protocol	connection protocol to be used	O
398.	applicationProfile	name of an application profile that can be used with the online resource	O
399.	name	name of the online resource	O
400.	description	detailed text description of what the online resource is/does	O
401.	function	code for function performed by the online resource	O

B.14.5 Series information

402.	CI_Series	information about the series, or aggregate dataset, to which a dataset belongs	Use obligation/condition from referencing object
403.	name	name of the series, or aggregate dataset, of which the dataset is a part	O
404.	issueIdentification	information identifying the issue of the series	O
405.	page	details on which pages of the publication the article was published	O

B.14.6 Telephone information

406.	CI_Telephone	telephone numbers for contacting the responsible individual or organization	Use obligation/condition from referencing object
407.	voice	telephone number by which individuals can speak to the responsible organization or individual	O
408.	facsimile	telephone number of a facsimile machine for the responsible organization or individual	O

B.15 Data structure for handling multi-languages support in free text metadata elements

	Name / Role name	Definition	Obligation / Condition
1x	PT_FreeText	description of a multi-language free text metadata element	Use obligation from referencing object
2x	Role name: textGroup	information about the metadata elements required to support multilingual free text fields	M
3x	PT_Group	description of metadata elements required to support multi-languages in free text metadata elements	Use obligation from referencing object
4x	languageCode	language used for documenting a plain text	O
5x	country	country of language used for documenting a plain text	O
6x	characterSetCode	full name of the ISO character coding standard used for documenting a plain text	O
7x	plainText	Content of a free text metadata element	M

Annex B reproduced by permission of *DIN Deutsches Institut für Normung e.V.* The definitive version for the implementation of this standard is the edition bearing the most recent date of issue, obtainable from *Beuth Verlag GmbH*, Burggrafenstrasse 6, D-10787 Berlin, Germany.

Annex C: Extensible Markup Language (XML)

The purpose of this annex is to give a short introduction to XML and the rationale behind the design of ISO 19118 (Encoding).

XML is an open, platform independent and vendor independent standard. It supports the international character set standards of ISO/IEC 10646 and Unicode. The XML standard is programming language neutral and API-neutral. A range of XML APIs are available, giving the programmer a choice of access methods to create, view, and integrate XML information. XML's tag structure and textual syntax make it as easy to read as HTML, and it is clearly better for conveying structured information. A growing set of tools is available for XML development.

General

The Extensible Markup Language (XML) is a subset of ISO 8879:1986 Standard Generalized Markup Language (SGML). XML has been designed for ease of implementation and for interoperability with both SGML and HTML. It is a format designed to bring structured information to the Web and in effect a language for electronic data interchange. XML is an open technology that is maintained by the World Wide Web Consortium (Bray et al. 1998).

XML defines a class of data objects called **XML documents**. A software module called an **XML processor** is used to read XML documents and provide access to their content and structure. An XML document is composed of pairs of tags, each pair consisting of a beginning and an end tag, like the <title></title> tags in HTML. Tag pairs may contain character strings and/or other tag pairs. A tag pair is called an element. In combination with its advanced linking capabilities, XML can encode a wide variety of information structures. The rules that specify how the tags are structured are called a Document Type Declaration or DTD. Alternately XML Schema can be used to give the rules. XML is in that respect similar to UML and has a declarative model (DTD) and an instance model (XML document). In HTML, the DTD Is pre-specified, with the writer allowed the use only of the tags specified in the HTML DTD, while in XML, writers may create their own DTDs with their own tag structure.

XML documents come in two flavours, i.e. **well-formed** XML documents and **valid** XML documents.

An XML document is well-formed if it conforms to the XML standard, and if it contains exactly one root element and any number of content elements where the elements, delimited by start- and end-tags, nest properly within each other.

A XML document is valid if it is well-formed and if it conforms to its DTD. An XML processor can be *non-validating* or *validating*. That means that a *non-validating* XML processor only checks if the XML document is well-formed. A *validating* XML processor checks whether the XML document conforms with its DTD as well.

XML element

The logical structure of an XML document consists of properly nested XML elements and entity references. An **XML element** may have attributes and a content. An element always has a start tag that may include the element's attributes, a content and an end tag. The content can be empty, a sequence of elements, one of an alternative list of elements, repetitions of elements, plain text or mixed elements and text data. An example of a simple XML element called **Road** with no attributes and plain text as content is:

```
<Road>Route 66</Road>
```

Here we see that "<Road>" is the start tag, "Route 66" is the content of the element, and "</Road>" is the closing tag of the element. A more complex XML element is an element with attributes and two elements as content:

```
<Person id="convenor1">
    <FirstName>John</FirstName>
    <LastName>Smith</LastName>
</Person>
```

Here we have an XML element called **Person** with an attribute **id** with value "convener1". The content of the **Person** element is two XML elements called **FirstName** and **LastName,** which have plain text as their content.

An XML element can refer to another element within the XML document or to external resources by using special purpose XML attributes. These XML elements are called linking elements. Linking elements point to a target resource through a Uniform Resource Identificator (URI) reference. Declaring XML elements with these special purpose XML attributes indicates their behaviour for the XML processor. See the next clause for an overview of some of the special purpose XML attributes. Here is an example of an XML element that points to another XML document:

```
<Reference xml:link="simple" href="http://www.example.org/AnnexA.xml">See Annex A</Reference>
```

Here the special purpose attribute **xml:link** indicates that this is a linking element and the value of the attribute **href** is a URI pointing to another XML document.

XML Schema

The DTD mechanism of XML has been criticised for lack of semantic expressiveness. The heavy uses of parameterised entities have made many DTDs difficult to read. XML Schema has been developed to solve that problem. An XML Schema is an XML document that defines the allowed elements and structures of XML documents. XML Schema enhances the type system of XML significantly. A type can be used to define the content model of an element. A type is declared using the type element:

```
<complexType name="RoadType" type="string"/>
```

Elements of this type will have text content model. Instead of having to use #PCDATA, we now can chose from a much richer set of types. This new type can then be used to declare an element:

```
<element name="Road" type="RoadType"/>
```

An element with a rich content model can be defined by a type with no name:
```
<element name="parent" >
```

Annex C: Extensible Markup Language (XML)

```
   <complexType>
      <sequence>
         <element name="c1" type="integer"/>
         <element name="c2" type="date" minOccurs="0"/>
         <element name="c3" type="RoadType" maxOccurs="*"/>
         <element name="c4" type="float" minOccurs="0" maxOccurs="*"/>
      </sequence>
   </complexType>
</element>
```

Here the parent element contains a sequence of four elements, where c1 shall appear once, c2 may appear, c3 may appear one or more times, and c4 may appear zero or more times.

An element with a choice of child elements is declared as follows:

```
<element name="car_part" >
   <complexType>
      <choice>
         <element ref="door"/>
         <element ref="wheel" />
         <element ref="engine" />
      </choice>
   </complexType>
</element>
```

Here the car_part element contains a group of elements that has a choice order. Thus, a car_part instance will contain one of the three elements, door, wheel or engine. Also notice the "ref" attribute of the element declaration. This refers to an element that is already defined somewhere else in the schema.

Attributes can also be specified in type definitions.

```
<complexType name="point">
   <attribute name="id" type="ID"/>
   <attribute name="dim" type="NMTOKEN" default="twoD">
      <enumeration value="oneD"/>
      <enumeration value="twoD"/>
      <enumeration value="threeD"/>
   </attribute>
   <attribute name="x" type="double"/>
   <attribute name="y" type="double"/>
   <attribute name="z" type="float" minOccurs="0"/>
</complexType>
```

Here we have defined five attributes for the type point. The "dim" attribute is of an enumerated type, where three values are allowed and "twoD" is the default value. The "z" attribute is optional.

Document Type Declaration (DTD)

A Document Type Declaration (**DTD**) declares the valid XML elements, their structure, and the XML entities that can be used by a class of XML documents. An XML document can have an external DTD subset defined in a separate file and/or an internal DTD subset defined as a part of the header information of the XML document. If an XML document contains both an external subset and an internal subset, the internal subset is considered to occur before the external subset. In the following DTD declarations are given with examples on what instances in the XML document will look like.

An XML element is declared using the special DTD start tag "<!ELEMENT":
<!ELEMENT Road (#PCDATA)>

```
<Road>Route 66</Road>
```

Here we define an XML element named **Road** with text as the content model. This is indicated by the XML reserved keyword "#PCDATA", which is short for parsed character data. The content model can be empty, a sequence or a choice of specific child elements, any combination of elements, or a mixture of text and specific child elements. A multiplicity operator can be used to specify the allowed occurrences of the child elements. The multiplicity operators are (?) for zero-or-one, (+) for one-or-more and (*) for zero-or-more occurrences. Absence of the multiplicity operator means that the child element must appear exactly once. Here is an example of the use of multiplicity operators:
<!ELEMENT parent (c1, c2?, c3+, c4*)>

```
<parent>
    <c1> ... </c1>
    <c3> ... </c3>
    <c3> ... </c3>
    <c4> ... </c4>
</parent>
```

The **parent** XML element shall contain the four child XML elements **c1, c2, c3** and **c4**, in that particular order. The element **c1** must always occur, **c2** is optional, there can be one-or-more **c3** elements and zero-or-more **c4** elements. The example shows one **c1**, no **c2**, two **c3** and one **c4** elements.

An XML element with a choice of child elements is declared as follows:
<!ELEMENT car_part (door | wheel | engine)>

```
<car_part>
    <wheel> ... </wheel>
</car_part>
```

Here a **car_part** element may contain a **door**, a **wheel** or an **engine**, but not more than one child element.

A mixed content model allows a mixture of text and specific child elements:
<!ELEMENT mp (#PCDATA | Road)*>

<mp>Here we can mix text and Road elements. <Road>E6</Road> See!</mp>

The ANY keyword specifies that the content model of an XML element can be anything:
<!ELEMENT p ANY>

<p>Anything goes <Road>E6</Road>here<tag/>!</p>

If an XML element is declared to be empty it shall not have a closing tag. This is indicated by a slash at the end of the start tag.
<!ELEMENT tag EMPTY>

An XML element can have attributes. The attributes are declared in an XML attribute list statement. XML has three groups of attribute types: character data, tokenised types and enumerated types. A #REQUIRED statement associated with the name of an attribute in the list means that it is mandatory; a #IMPLIED statement implies that it is optional. It is also possible to give default values for attributes. An attribute is declared in an attribute list construct:

```
<!ELEMENT point EMPTY>
<!ATTLIST point
          id            ID                       #REQUIRED
          dim           (oneD | twoD | threeD)   #IMPLIED "twoD"
          x             CDATA                    #REQUIRED
          y             CDATA                    #IMPLIED
          z             CDATA                    #IMPLIED>
```

Here we have defined an XML element named **point**. The **point** has an attribute list of five attributes. The attribute called **id** has a tokenised type ID and it is required. The **dim** attribute has an enumerated type that takes one out of three valid values. This attribute is implied, but it has a default value of "twoD". The **x**, **y** and **z** attributes all are of type "CDATA", which is short for character data. Since XML does not have any data types for numeric values the coordinate values must be converted to strings. Notice that only **x** is required. Two instances of this XML element are:

<point id="i01" x="12300" y="234"/>
<point id="i02" dim="threeD" x="123456" y="-234567" z="50"/>

For the first instance with **id** equal "i01" we can use the implied default value of **dim** to indicate that this point is two dimensional and do not have to give this explicitly. The second instance has all attributes filled in.

The tokenised attribute types has special restrictions on the allowed characters and they are:

Table C.1. DTD attribute types

XML Type	Semantics
ID	An identifier for the element that, if specified, must be unique within the XML document. The value of the identifier must always start with a letter, "_" or ":". An XML element can only have one attribute of type ID.
IDREF	A reference to an XML element in the XML document. The value must correspond to an attribute value of type ID in an existing XML element.
IDREFS	A reference to one or more XML elements. The values must be separated by spaces and must correspond to existing XML element ID's.
ENTITY	A reference to an external entity. The value must be a legal entity name.
ENTITIES	A reference to any number of entity names, where the entity names are separated by spaces.
NMTOKEN	A NMTOKEN (Name Token) is any mixture of characters.
NMTOKENS	Any number of NMTOKENs separated by spaces.

Some attributes are specific for XML and have reserved names and well-defined semantics. The **xml:lang** attribute can be used to indicate the language used in XML elements and **xml:space** can be used to indicate how to handle whitespace within elements.

Table C.2. Two special purpose XML attributes

Attribute name	XML Type	Semantics
xml:lang	NMTOKEN	A special attribute that may be inserted in documents to specify the language used in the contents and attribute values of any element in an XML document. But it must be defined in the attribute list specification of the actual element.
xml:space	(default | preserve)	A special attribute that signals an intention that in that element, white space should be preserved by applications.

Linking element

Linking elements are recognized based on the use of a designated attribute named **xml:link** and a set of accompanying attributes. This is described in the XML XLink specification (XML00 2000). A **link** is an explicit relationship between two or more data objects or portions of data objects often represented by a URI or an IDREF. The content of the linking element is called the **local resource** and the target of the link is called the **remote resource**. A remote resource is identified by a text string called a **locator**. A locator value may contain either a Uniform Resource Identifier (URI) or a fragment identifier, or both. The syntax of a locator is first the URI, followed by a **connector** ("#" or "|") and a fragment identifier. The URI describes a remote resource, and the fragment identifier describes a sub-resource within that resource. A **fragment identifier** pointing into an XML document must be an Xpointer (XML01 2001). If the connector is "#", this signals that the remote resource is to be fetched as a whole, and that the XPointer processing to extract the sub-resource is to be performed on the client. If the connector is "|", no intent is signalled as to what processing model is to be used for accessing the designated resource.

The following information can be associated with a link and its resources: One or more locators to identify the remote resources participating in the link (a locator is required for each remote resource), the type of the link, the type of the remote resources and the type of the local resource. Example of a linking element is:

    <link xml:link="simple" href="http://www.example.org/data.xml|id(i005)" >This is the local resource</link>

Here we have a linking element called **link**. It has two attributes **xml:link** and **href**. **xml:link** states that this is a simple linking element, whereas **href** holds the locator identifying the remote resource. The locator consist of a URI which is "http://www.example.org/data.xml", a connector "|" and a fragment identifier "id(i005)" in XPointer syntax. We could also have used "i005" directly, it is defined as a shortcut in the XPointer specification. The content of the linking element is the local resource. Other combinations can be:

```xml
<link xml:link="simple" href="|id500">This link points to an element within the document with an attribute of type ID with value "id500"</link>

<link xml:link="simple" href="xdata.xml">This link points to an XML document</link>
```

The XLink specification defines the following attributes:

Table C.3. XLink attributes

Attribute name	XML Type	Semantics
xml:link	CDATA	This is a special reserved attribute that indicates that the element shall act as a linking element. Legal values are: simple, extended, locator, group or document. We will only use simple links!
href	CDATA	The value of the href attribute in linking elements contains a locator which identifies a resource, e.g. by a URI-reference or by an XPointer specification.
inline	(true | false)	The inline attribute specifies the first part of a link's semantics. A link is either inline or out-of-line. This attribute is used in connection to the extended links. Inline is within the document, whereas out-of-line is outside the document.
role	CDATA	The role attribute also specifies a part of the link's semantics. The value of this attribute identifies to the application software the meaning of the link. This allows the application to show different symbols for the different kinds of links.
title	CDATA	This is the title shown to the user for the remote resource.
show	(embed | replace | new)	This attribute indicates the behaviour policies to use when the link is traversed for the purpose of display or processing. The embed value indicates that the designated resource should be embedded in the body of the resource and at the location where the traversal started. The replace value indicates that the designated resource should replace the resource where the traversal started. The new value indicates that the designated resource should be displayed or processed in a new context.
actuate	(auto | user)	The actuate attribute is used to express a policy as to when the traversal of a link should occur. The auto value indicates that the resource is automatically traversed. The user value indicates that the link is traversed only on the request of the user.

XML entity

An XML entity is a mechanism for text substitution. It can make XML documents simpler to read and quicker to write. In the following we define an XML entity named "XML" which are used in a "p" element:

```
<!ENTITY XML "Extensible Markup Lanuage">
<!ELEMENT p (#PCDATA) >

    <p>This is written in XML (&XML;).</p>
```

The entity reference "&XML;" in the <p> element indicates that the reference shall be replaced with the text of the entity. Only text entities that can be parsed as XML can be referenced directly in the XML document.

XML entities are divided into text and binary entities. Binary entities do not contain valid XML and must therefore be referenced in an XML element's attribute of the ENTITY type. Here we define an external binary entity named **my.sign** and refer to it in the **src** attribute of an **image** element.

```
<!ENTITY my.sign SYSTEM "image/signature.gif" NDATA GIF>
<!ELEMENT pi ANY>
<!ELEMENT image EMPTY>
<!ATTLIST image
            src           ENTITY         #REQUIRED >

    <pi>This is my signature: <image src="my.sign"/></pi>
```

There is also a distinction between internal entities and external entities. An internal entity has a value associated with it as part of the entity declaration. An external entity associates a name with a physical storage unit (file name). The "my.sign" entity above is an example of an external entity and the "XML" entity is an example of an internal entity.

There is a special construct that only can be used in the DTD called a parameter entity. A **parameter entity** can be used as a shortcut for commonly occurring text structures. A parameter entity declaration is similar to an entity declaration except with a "%" sign followed by a space before the entity name. The parameter entity will be substituted with its text by the XML processor when it reads the DTD. An example of a declaration of a parameter entity called "Boolean" and its usage is as follows:

```
<!ENTITY % Boolean "( true | false )" >

<!ELEMENT my_Boolean_element EMPTY>
<!ATTLIST my_Boolean_element %Boolean; >
```

Character coding

Each XML text entity must declare the character-encoding scheme that it uses internally. External parsed entities in an XML document may use different encoding schemes for their characters than used in the root document. All XML processors must accept the UTF-8 and UTF-16 encoding of ISO/IEC 10646 as a minimum. Parsed entities, which are stored in an encoding other than UTF-8 or UTF-16 must begin with an encoding declaration in the document entity:

<?xml version="1.0" encoding="ISO-10646-UCS-2" ?>

According to the XML recommendation the following values are allowed:

- For ISO/IEC 10646 and UNICODE based encoding "UTF-8", "UTF-16", "ISO-10646-UCS-2" and "ISO-10646-UCS-4" shall be used. If an XML text entity is encoded in UCS-2, it must start with an appropriate encoding signature, the Byte Order Mark, which is the character with hexadecimal value FEFF.
- For ISO/IEC 8859: "ISO-8859-1", "ISO-8859-2", ... , "ISO-8859-10" can be used.
- For various encodings of JIS X-0208-1997: "ISO-2022-JP", "Shift_JIS" and "EUC-JP" can be used.

This is restricted to "UTF-8", "UTF-16", "ISO-10646-UCS-2" and "ISO-10646-UCS-4" in this International Standard.

Alternatively one can reference any character in ISO/IEC 10646 by quoting its character number in a character reference regardless of which encoding scheme used. Or one can declare a text entity that represents the character in question. There are two ways of referring to characters directly, by decimal representation or by hexadecimal reference. The hexadecimal reference for the less-than-sign "<" is:

<

And the decimal representation of the same sign is:

<

XML document header

All XML documents must start with a processing instruction that specifies that this is an XML document and which version of the XML standard is used. Information about the character encoding, if it is not "UTF-8" or "UTF-16", shall be included in the header also. An example is as follows:

<?xml version="1.0" encoding="ISO-10646-UCS-2" ?>.

The next element shall be the document type declaration element:

```
<!DOCTYPE top SYSTEM "root.dtd" [
  <!ELEMENT bot EMPTY> ] >
```

Here we see a document type element declaration that has both an external subset and an internal subset DTD. An external subset DTD is located in a separate file, whereas a internal subset DTD is included in the XML document. The root XML element is called **top** and it must be defined in the external DTD subset declared in the "root.dtd" file. The internal DTD subset is declaring a single XML element called **bot**.

Miscellaneous

An XML comment may appear anywhere in the document, but outside other markup. Comments are not part of the document's character data. A comment starts with the string "<!--", followed by any characters except the string "--" and ends with the string "-->". An example of a comment is:

```
<!-- This is a comment. -->
```

All of an XML document is case sensitive, both markup and text. This is different from SGML and HTML and it was introduced to allow markup in non-Latin alphabet characters and to avoid the problem with case folding. This means that element names, entity names and attribute names all are case sensitive. The following elements are different and thus allowed in an XML document:

```
<road>This is a road element</road>
<Road>This is a Road with a capital R</Road>
```

Other XML standards

The base XML standard is associated with a number of supporting standards. The most relevant standards are shortly described in this clause.

A **link** is an explicit relationship between two or more data objects or portions of data objects. In addition to the IDREF mechanism in XML documents a link can be constructed with attributes on XML elements using a combination of the XLink and XPointer specification. The **XML Linking Language** and the **XML Pointer Language** specify constructs to define links between both external and internal objects of XML documents. XLink is used to create simple unidirectional hyperlinks between objects in separate XML documents as well as more sophisticated links. Whereas XPointer is used for addressing internal structures within XML documents, such as references to elements, character strings and other parts of XML documents.

The **Namespace in XML** specification (XML99 1999) provide a simple method for qualifying element and attribute names used in XML documents by associating them with a namespace identified by a URI reference. An **XML namespace** is a collection of names, identified by a URI reference, which are used in XML documents as element types and attribute names. XML documents can then mix XML elements and attributes from more than one DTD that may have identical names and different semantics in the different DTDs. The mechanism for achieving this is to use qualified names for both XML elements and attributes. A **qualified name** consists of a namespace prefix, a single colon as separator, followed by the local name. Here is an example of the use of the XML namespace mechanism:

```
<x xmlns:gis="http://www.example.org/schema/spatial.dtd">
    <gis:point>234 445 150</gis:point>
    <point>23 55</point>
</x>
```

Here we see **x** contain two elements named point. The local **point** element is defined in the DTD of the XML document, but the **gis:point** element is defined in a DTD located at URI "http://www.example.org/shema/spatial.dtd". The special purpose attribute **xmlns:gis** in element **x** defines **gis** as a namespace prefix of the declarations defined in the target DTD.

The **Resource Description Framework** (RDF) is a result of the W3C Metadata Activity. RDF is the foundation for processing and exchanging machine-understandable metadata on the Web using XML as the exchange format. It can be used in a variety of application areas: in resource discovery, in cataloguing for describing the content and content relationships available at a particular Web site, in content rating, in describing collections of pages that represent a single logical document, for describing intellectual property rights of Web pages, and for expressing the privacy preferences of a user as well as the privacy policies of a Web site. The **RDF Model and Syntax Specification** (RDF99 1999) document defines a data model for representing named properties and property values. The data model consists of three object types: Resource, Property and Statement. Instances of the RDF data model are called a RDF Schema. RDF resources are things of interest, e.g. an entire Web page, an XML element or an entire Web site. An RDF property is a specific aspect, characteristic, attribute or relation used to describe a resource, whereas an RDF statement is a specific RDF resource in combination with a named RDF property for that resource. RDF can be used to express a wide variety of data models, e.g. Entity-Relationship models. The **RDF Schema Specification** (RDF00 2000) document defines a schema specification language that provides a basic type system for use in RDF models. It defines resources and properties such as class and subclass-of constructs that can be used in application specific schemas. RDF Schemas can be compared with XML DTDs, but unlike an XML DTD, which gives specific constraints on the structure of an XML document, a RDF Schema provides information about the interpretation of the statements given in an RDF data model.

XMI (XMI03 2003) is an XML based exchange standard for exchange of object-oriented metadata models. The purpose of XMI is to facilitate UML model data interchange between different modelling tools in a vendor neutral way. It is based on OMG's Meta Object Facility and on CORBA data types. XMI can in theory be used to exchange data based on UML models, but are not primarily designed for this purpose.

The **Extensible Stylesheet Language** (XSL) (XSL 1999) is intended to control the appearance of XML documents. XSL is a language for expressing stylesheets. A stylesheet expresses rules for presenting a class of XML documents. Thus a stylesheet contains descriptions on how XML elements can be rendered by an XML browser. XSL views an XML document as a tree and uses a two-stage presentation process. First, the result tree is constructed from the source tree. Second, the result tree is interpreted to produce formatted output on display, on paper, in speech or onto other media.

The **Document Object Model** (DOM 1998) specification defines a platform- and language-neutral interface, that allows programs and scripts to dynamically access and update the content, structure and style of XML documents. The DOM provides a standard set of objects for representing HTML and XML documents, a standard model of how these objects can be combined, and a standard interface for accessing and manipulating them. This means that vendors can support DOM as an interface to their proprietary XML processors.

The **Scalable Vector Graphics** language (SVG 2000) is a language for describing two-dimensional graphics in XML. SVG allows for three types of graphic objects: vector graphic, images and text. Graphical objects can be grouped, styled, transformed and composed into previously rendered objects. SVG drawings can be interactive and dynamic. Animations can be defined and triggered either declaratively (i.e., embedding SVG animation elements in SVG content) or via scripting.

Annex C reproduced by permission of *DIN Deutsches Institut für Normung e.V.* The definitive version for the implementation of this standard is the edition bearing the most recent date of issue, obtainable from *Beuth Verlag GmbH*, Burggrafenstrasse 6, D-10787 Berlin, Germany.

Annex D: Abbreviations

Acronyms

AdV	Arbeitsgemeinschaft der Vermessungsverwaltungen der Länder der Bundesrepublik Deutschland (Working Committee of the Surveying Authorities of the States of the Fereral Republic of Germany)
AFIS	Amtliches Festpunkt-Informationssystem (Authoritative Geodetic Control Point Information System)
AFNOR	Association française de normalisation
AG	Advisory Group
AGI	Association of Geographic Information
ALB	Automatisiertes Liegenschaftsbuch (Automated Real Estate Book)
ALK	Automatisierte Liegenschaftskarte (Automated Cadastral Map)
ALKIS	Amtliches Liegenschaftskataster-Informationssystem (Authoritative Cadastral Map Information System)

ANSI	American National Standards Institute
API	Application Programming Interface
ARCS	Admiralty Raster Chart Standard
ASCII	American Standard Code for Information Interchange
ASPRS	American Society for Photogrammetry & Remote Sensing
ATKIS	Amtliches Topographisch-Kartographisches Informationssystem (Authoritative Topographic Cartographic Information System)
ATS	Abstract Test Suite
AVHRR	Advanced Very High Resolution Radiometer
AWAR [lp/mm]	Area weighted average resolution
BAE	British Aerospace (and Marconi) Electronic
BIIF	Basic Image Interchange Format
BNSC	British National Space Center
BOD	Board of Directors (OGC)
C/A	Coarse / Acquisition code transmissions of the GPS and GLONASS
C++	Programming language based on C with object-oriented extensions
CCRS	Compound Coordinate Reference System
CCRS	Canada Centre for Remote Sensing
CD	Committee Draft
CEN	Comité Européen de Normalisation
CEO	Center of Earth Observation
CEO	Chief Executive Officer
CEOS	Committee on Earth Observation Satellites
CGI	Computer Graphics Interface
CGM	Computer Graphic Metafile
CHRIS	Committee on Hydrographic Requirements for Information Systems (IHO)
CHS	Canadian Hydrographic Service
CIE	Commission internationale de l'eclairage
CIG	Canadian Institute of Geomatics
CLA	Cultural and Linguistic Adaptability
CNES	Centre National d'Etudes Spatiales
COM	Component Object Model
CORBA	Common Object Request Broker Architecture
COTS	Commercial Off-The-Shelf
CP IDEA	Comité Permanente para la Infraestructura de Datos Especiales de las Américas
CR	CEN Report
CRS	Coordinate Reference System
CRSS	Canadian Remote Sensing Society
CSA	Canadian Space Agency
CSL	Conceptual Schema Language
CSMF	Conceptual Schema Modelling Facility
DCP	Distributed Computing Platform
DE	Digital Earth

Annex D: Abbreviations

DEM	Digital Elevation Model
DG	Directorate-General
DGI	Digital geographic information
DGIWG	Digital Geographic Information Working Group (NATO)
DGPS	Differential GPS
DGPF	Deutsche Gesellschaft für Photogrammetrie und Fernerkundung
DIGEST	Digital Geographic Information Exchange Standard
DIS	Draft International Standard
DIN	Deutsches Institut für Normung e.V.
DLR	Deutsches Zentrum für Luft- und Raumfahrt
DNC	Digital Nautical Chart (DIGEST)
DOM	Document Object Model
DOP	Dilution of Precision
DTD	Document Type Definition
ECDIS	Electronic Chart Display and Information System (S-57)
ECI	Earth Centered Inertial Coordinate Reference Systems
ECS	EOSDIS Core System
EDOC	Enterprise Distributed Object Computing
EGM	Earth Gravity Model
EJB	Enterprise Java Beans
ENC	Electronic Nautical Chart (S-57)
ENV	European pre-Standard
EOS	Earth Observation System
EOSE	Extended Open Systems Environment Model
EOSDIS	EOS Data Information System
EPSG	European Petroleum Survey Group
ERS	European Remote Sensing Satellite
ESA	European Space Agency
ESRI	Environmental Systems Research Institute, Inc.
ETS	Executable Test Suite
EUROGI	European Umbrella Organization for Geographic Information
FACC	Feature and Attribute Coding Catalogue (DIGEST)
FAO	Food and Agriculture Organization
FDDI	Fibre Distributed Data Interface
FDIS	Final Draft International Standard
FGDC	(U.S.) Federal Geographic Data Committee
FIG	International Federation of Surveyors
FOM	Figure of Merit
FRS	Feature Representation Scheme (GDF)
FTP	File Transfer Protocol
GDF	Geographic Data Files (CEN/ISO)
GDIN	Global Disaster Information Network
GDOP	Geometric Dilution of Precision
GIF	Graphics Interchange Format
GILS	Government Information Locator Service
GIOP	General Inter-ORB Protocol

GIS	Geographic Information System
GFM	General Feature Model
GML	Geography Markup Language
GKS	Graphical Kernel System
GLONASS	GLObal NAvigation Satellite System (Russian Federation)
GNSS	Global Navigation Satellite System (generic)
GPS	Global Positioning System (USA)
GRASS	Geographic Resources Analysis Support System
GRSS	IEEE Geoscience and Remote Sensing Society
GSDI	Global Spatial Data Infrastructure
HDF	Hierarchical Data Format
HDOP	Horizontal Dilution of Precision
HIS	Information Technology Human Interaction Service
HMMG	Harmonized Model Maintenance Group
HTI	Human Technology Interface
HTML	Hypertext Markup Language
HTTP	Hypertext Transfer Protocol
HWP	Harmonization Working Party (joint DGIWG/IHO)
IAG	International Association of Geodesy
IBM	International Business Machines
ICA	International Cartographic Association
ICAO	International Civil Aviation Organization
ICAN	International Commission on Air Navigation
ICD	Interface Control Document (DGIWG/IHO HWP)
ICS	Implementation Conformance Statement
ICS	Interoperable Catalogue System
ID	Identifier
IDL	Interface Definition Language
IDREF	An XML ID reference type
IEC	International Electrotechnical Commission
IGARSS	International Geoscience and Remote Sensing Symposium
IHB	International Hydrographic Bureau (secretariat of the IHO)
IHO	International Hydrographic Organization
IIOP	Internet Inter-ORB Protocol
IMO	International Maritime Organization
InSAR	Interferometric SAR (Synthetic Aperture Radar)
INSPIRE	Infrastructure for Spatial Information in Europe Initiative
IPI	Image Processing and Interchange
IRDS	Information Resource Dictionary System
IS	International Standard
ISA	International Federation of the National Standardising Associations
ISCGM	International Steering Committee for Global Mapping
ISO	International Organization for Standardisation ("iso" = Greek for "same")
ISP	International Standardised Profile
ISPRS	International Society for Photogrammetry & Remote Sensing

Annex D: Abbreviations

IT	Information Technology
ITRF	International Terrestrial Reference Frame
ITRS	International Terrestrial Reference System
ITS	Intelligent Transportation System
ITU	International Telecommunication Union
ITU-T	ITU Telecommunication Standardisation Sector
IUT	Implementation Under Test
IWA	International Workshop Agreement
IXIT	Implementation extra information for testing
J2EE	Java 2 Enterprise Edition with EJB
JBIG	Joint Binary Images Group
JDBC	Java Data Base Connectivity
JERS	Japanese Earth Resources Satellite
JFIF	JPEG File Interchange Format
JIM	Java IDL
JPEG	Joint Photographic Experts Group
JPL	Jet Propulsion Laboratory (NASA)
JRC	Joint Research Centre of the European Union
JSP	Java Server Pages
JTC1	Joint Technical Committee 1
LANDSAT	Land Satellite (U.S.)
LBS	Location Based Services
Lidar	Light detecting and ranging
LISP	Programming language based on LISt Processing. The standard is called Common LISP.
LORAN-C	LOcation and RANging radiolocation system
LRS	Linear Reference System
LUT	Look-Up-Table
MBR	Minimum Bounding Region
MC	Management Committee (OGC)
MDA	Model Driven Architecture
MHEG	Multimedia/Hypermedia information Coding Experts Group
MIME	Multipurpose Internet Mail Extensions
MPEG	Moving Pictures Expert Group
MODIS	Moderate Resolution Imaging Spectroradiometer (NASA)
MOS	Marine Observation Satellite (Japan)
MTPE	Mission To Planet Earth
MTS	Microsoft Transaction Server
NADyy	North American Datum; suffix "yy" indicates last two digits of year
NAS	Normenbasierte Austauschschnittstelle (Standard-based data exchange interface)
NASA	National Aeronautics and Space Administration (U.S.)
NASDA	National Space And Development Agency of Japan
NATO	North Atlantic Treaty Organization
NCGIA	National Center for Geographic Information & Analysis

NIMA	National Imagery and Mapping Agency (U.S.)
NISO	National Information Standards Organization
NIST	National Institute of Standards and Technology
NITF	National Imagery Transmission Format (U.S.)
NMEA	National Marine Electronics Association
NOAA	National Oceanic and Atmospheric Administration (U.S.)
NSDI	National Spatial Data Infrastructure (U.S.)
NSIF	NATO Secondary Imagery Format
NWIP	New Work Item Proposal
OCL	Object Constraint Language
OCR	Optical Character Recognition
ODBC	Open Database Connectivity
ODP	Open Distributed Processing (see RM-ODP)
OGC	Open GIS Consortium, Inc.
OGIS	Open GIS
OLE	Object Linking and Embedding
OMG	Object Management Group
ORB	Object Request Broker
OSE	Open Systems Environment
OSI	Open Systems Interconnection
OWS	OGC Web Service
PAS	Publicly Available Specification (ISO)
PCGIAP	Permanent Committee on GIS Infrastructure for Asia and the Pacific
PC IDEA	Permanent Committee on Spatial Data Infrastructure for the Americas
PCL	Printer Control Language
PDA	Personal Digital Assistant
PDF	Portable Document Format (Adobe)
PHIGS	Programmer's Hierarchical Interactive Graphic System
PIKS	Programmer's Imaging Kernel System
PIXEL	Picture Element
PNG	Portable Network Graphics
PDOP	Positional Dilution of Precision
PPS	Precise Positioning Service of a Global Navigation Satellite System
PT	Project Team
PTT	Protocol Task Team
RADARSAT	Radar Satellite (Canada)
RAIM	Receiver Autonomous Integrity Monitoring
REPMAT	Reproduction Material
RFC	Request for Comments
RFI	Request for Information
RFP	Request For Proposal
RFT	Request for Technology
RGB	red, green, blue

RINEX	Receiver INdependent EXchange format
RNCP	Raster Nautical Chart Products
RM-ODP	Reference Model of Open Distributed Processing (ISO/IEC 10746)
RMS	Root Mean Square
RMSE	Root Mean Square Error
S-57	IHO Transfer Standard
SAR	Synthetic Aperture Radar
SC	Subcommittee
SCAR	Scientific Committee on Antarctic Research
SCC	Standardisation Council of Canada
SCSI	Small Computers Systems Interface
SDI	Spatial Data Infrastructure
SDO	Standards Developing Organization
SDTS	Spatial Data Transfer Standard
SGML	Standard Generalized Markup Language
SI	Le Système International d'Unités
SIG	Special Interest Group (OGC)
SLD	Styled Layer Descriptor
SLR	Single Lens Reflex camera
SMIL	Synchronized Multimedia Integration Language
SNR	Signal to Noise Ratio
SOAP	Simple Object Access Protocol
SOF	Service Organizer Folder
SPOT	Sytème Pour l'Observation de la Terre (France)
SQL	Structured Query Language
SQL/MM	SQL/ Multi-Media
SQL 99	The SQL language specification adopted in 1999, which includes object-oriented data-type extension mechanisms
SRI	Stanford Research Institute (SRI International)
SRPE	SDTS Raster Profile and Extension
SSM/I	Special Sensor Microwave Imager (NASA)
STANAG	Standardisation Agreement. (NATO)
STEP	Standard for Exchange of Product Data
SUT	System Under Test
SV	Space Vehicle
SVG	Scalable Vector Graphics
TC	Technical Committee
TDOP	Time Dilution of Precision
TF	Task Force
TICS	Transport Information and Control System
TIFF	Tag Image File Format
TIFF/IT	Tag Image File Format for Image Technology
TIN	Triangulated Irregular Network
TMAG	Terminology Maintenance Advisory Group
TMB	Technical Management Board
TMG	Terminology Maintenance Group

TR	Technical Report
TS	Technical Specification
TSMAD	Transfer Standard Maintenance and Applications Development Working Group (IHO)
UCGIS	University Consortium for Geographic Information Science
UCS	Universal Multiple-Octet Coded Character Set
UDT	User Defined Type
UNIGIS	University Consortium for Certificate & Graduate Programs in GIS
URISA	Urban & Regional Information System Associates
UML	Unified Modelling Language
UN	United Nations
UNECE	United Nations Economical Commission for Europe
UNEP	United Nations Environment Programme
UNGEGN	United Nations Group of Experts on Geographical Names
UNGIWG	United Nations Geographic Information Working Group
UOD	Universe of Discourse
UOM	Units of measurement
URI	Uniform Resource Identifier (other name: URL)
URL	Uniform Resource Locator
USGS	United States Geological Survey
USOC	Use of the Object Catalogue for ENC (S-57)
UTC	Coordinated Universal Time
UTF	UCS Transfer Format
UTM	Universal Transverse Mercator
UUID	Universal Unique Identifier
VDOP	Vertical Dilution of Precision
VRF	Vector Relational Format (DIGEST encapsulation)
VRML	Virtual Reality Modelling Language
W3C	World Wide Web consortium
WebCGM	Web Computer Graphics Metafile
WCS	Web Coverage Service
WD	Working Draft
WFS	Web Feature Server
WG	Working Group
WGISS	Working Group on Information Systems and Services (CEOS)
WMO	World Meteorological Organization
WMS	Web Map Service
X3D	WEB3D Consortium
XMI	XML Metamodel Interchange
XML	Extensible Markup Language
XSD	XML Schema Document
XSL	Extensible Stylesheet Language
Xpointer	XML Pointer language

Mathematical symbols

a	semi-major axis
a	ground range resolution (x direction) (SAR sensors)
b	semi-minor axis
b	azimuth resolution (y direction) (SAR sensors)
β	depression angle (SAR sensors)
c' [mm]	calibrated focal length
e	slant range resolution (SAR sensors)
E	easting
f	flattening
h	ellipsoidal height
H	gravity-related-height
H_n	flying height (SAR sensors)
λ	geodetic longitude
O'	projection centre, X_O, Y_O, Z_O [m], coordinates of O'
PPA', x_0', y_0' [mm]	principal point of autocollimation
φ	geodetic latitude
R_g	ground range (SAR sensors)
R_s	slant range (SAR sensors)
φ [deg]	Rotation around y-axis
Φ_n	near edge incidence angle (SAR sensors)
Φ_f	far edge incidence angle (SAR sensors)
2D	Two-dimensional
3D	Three-dimensional

Annex E: Class names

The class names in the UML-diagrams of the 19100-series of standards begin with a two-character-acronym for a better understanding of the classes and ther inter-relations. The two-character-acronym are defined by the Harmonized Model Maintenance Group (HMMG) of the ISO/TC211.

AD	Abstract address	ISO 19133
CC	Changing coordinates	ISO 19111
CI	Citation	ISO 19115
CV	Coverages	ISO 19123
DQ	Data quality	ISO 19115
DS	Dataset	ISO 19115
EX	Extent	ISO 19115
FC	Feature Catalogue	ISO 19110
FD	Feature ID	ISO 19133
FE	Feature	ISO 19109

FT	Feature Topology	ISO 19107
GF	General feature model	ISO 19109
GM	Geometry model	ISO 19107
GR	Graph	ISO 19107
LI	Lineage	ISO 19115
LR	Linear reference	ISO 19133
MD	Metadata	ISO 19115
NT	Network position	ISO 19133
PF	Feature Portrayal	ISO 19117
PS	Positioning Services	ISO 19116
RS	Reference System	ISO, 19111, ISO 19115
SC	Spatial coordinates	ISO 19111
SD	Sensor and Data Models	ISO 19130
SI	Spatial Identification	ISO 19112
SV	Services	ISO 19119
TM	Temporal	ISO 19108
TP	Topology	ISO 19107
TS	Simple Topology	ISO 19107

Annex F: Past and planned meetings

1st Plenary	November 1994	Oslo, Norway
2nd Plenary	August 1995	Reston, Virginia, USA
3rd Plenary	May 1996	Seoul, Republic of Korea
4th Plenary	January 1997	Sydney, Australia
5th Plenary	October 1997	Oxford, UK
6th Plenary	March 1998	Victoria, British Columbia, Canada
7th Plenary	September 1998	Beijing, China
8th Plenary	March 1999	Vienna, Austria
9th Plenary	September 1999	Kyoto, Japan
10th Plenary	March 2000	Cape Town, South Africa
11th Plenary	September 2000	Reston, Virginia, USA
12th Plenary	March 2001	Lisbon, Portugal
13th Plenary	October 2001	Adelaide, Australia
14th Plenary	May 2002	Bangkok, Thailand
15th Plenary	November 2002	Gyeongju, Republic of Korea
16th Plenary	May 2003	Thun, Switzerland
17th Plenary	October 2003	Berlin, Germany
18th Plenary	spring 2004	Kuala Lumpur, Malaysia
19th Plenary	fall 2004	JRC (European Commission), Ispra, Italy
20th Plenary	spring 2005	Stockholm, Sweden
21st Plenary	fall 2005	Riyadh, Saudi Arabia
22nd Plenary	spring 2006	USA

Index

A

Abbreviations 299-307
Abstract Specification 110, 179, 183, 184, 209, 210
accreditation 123
Advisory Group 31-33, 51, 52
AFNOR 29
aggregation 59, 62, 131, 157, 172, 179, 215
AGI 124, 299
Aircraft and space vehicles 12, 186, 187
ALKIS 49
American National Standards Institute (ANSI) 10, 92, 101, 300
ANSI 10, 92, 101, 300
application model 43-46
Application Programming Interface (API) 73, 74
application schema 30, 36, 37, 41, 44, 48-50, 53, 55, 68, 71, 75, 76, 87-89, 94, 113, 177, 181-184, 214, 215, 227, 280
Approval stage 13, 14, 18
arc (Arc) 126, 132, 133-135, 143, 155, 156, 178
arc string (ArcString) 131, 134, 135, 178
arc string by bulge (ArcStringByBulge) 135, 136, 178
architectural reference model 45, 47, 72, 73
area 131, 153, 157, 162, 169
asGM_Geodesic 154, 155
asGM_LineSegment 154
association 56, 57, 59, 62, 63, 76, 88, 113, 114, 123, 124, 182, 215

Association of Geographic Information 124, 299
attribute 37, 57, 75, 88, 126, 166-168, 172, 175-177, 180, 182

B

BAE Systems 207, 209, 300
base standards 30, 32, 33, 50, 70, 227
Basic Image Interchange Format (BIIF) 29, 101, 191, 300
basic test 67, 227
bearing 153, 155, 156
Bernstein polynomials 139, 140
Bézier (Bezier) 128, 140, 178
bicubic grid 132, 144
BIIF 29, 101, 191, 300
bilinear grid 132, 143
Bluetooth 117
Board of Directors (BOD) 209, 210, 300
Booch 56
boundary 126, 128, 130, 140, 145, 147, 153, 157, 162, 176, 179, 180
Boustrophedonic scanning 173
B-spline (BSpline) 137, 139, 140, 178
B-spline surface 132, 144
buffer 150, 227

C

cadastre 27, 29, 50, 52, 130, 159, 204, 214-216
Cadcorp 209
Cantor-diagonal scanning 173
capability test 67, 227
Cartesian system 77-81, 228

cartography 29, 85, 159, 198, 203, 204
Cascading Web Map server 117, 213, 214
cataloguing (catalogue) 38, 52, 89, 111-114, 213-215
CD 13, 15, 212, 300
cellular phone 117
CEN 6, 29, 30, 32, 34, 214, 300
centroid 149, 150
CEOS 101, 192-194, 300, 306
certification of personnel 123, 124
CGI 28, 29, 191, 300
CGM 28, 29, 191, 300
character coding 296
circle (Circle) 132, 135, 137, 142, 148, 155, 156, 178
city maps 27
class-diagram 38, 56, 59-64, 129
clearinghouse 89, 211
closure 147, 228
clothoid 131-133, 137, 178
Coded character sets (JTC1/SC2) 186, 190
Coding of audio, picture, multimedia and hypermedia information (JTC1/SC29) 186, 191
Comité Européen de Normalisation (CEN) 6, 29, 30, 32, 34, 214, 300
Committee Draft (CD) 13, 15, 212, 300
Committee on Earth Observation Satellites 101, 192-194, 306, 300
Committee stage 13, 15, 18
common architecture 55, 115, 125
Communication Service (ITCS) 73, 74
Communications Service Interface (CSI) 73, 74
completeness 37, 94
composition 59, 62, 166
Compusult 213, 214
computational viewpoint 42, 43, 72
Computer graphics and image processing (JTC1/SC24) 186, 190, 191

Computer Graphics Interface 28, 29, 191, 300
Computer Graphics Metafile 28, 29, 191, 300
conceptual model 44-46, 230
conceptual schema language 44, 46, 55, 56, 214, 230
Conceptual Schema Modelling Facilities (CSMF) 43, 45, 300
conceptualisation principle 45
cone 132, 142
conformance 8, 16, 17, 21, 32, 53, 55, 67, 68, 70, 95, 184, 230, 245, 249
conformance and testing 16, 32, 53, 55, **67, 68**
conic 131, 136, 137, 142, 178
conic scanners 109
consortium 1, 5, 6, 8, 9, 11, 31, 33, 48, 51, 67, 110, 115-117, 124, 181, 183, 184, 193, 197, 202, 207-212, 213, 215, 216
constraint 56, 57, 59, 61, 62, 76, 109, 114, 122, 130, 131, 215, 264
contains 151
content model 183, 288, 290, 291
continuous coverage 169-177, 181
conversion 79, 84, 89, 133, 186, 187
convexHull 150, 231
Coordinate Reference System 8, 32, 39, 40, 44, 72, **77-81,** 97, 99, 106, 109, 116, 120, 130, 134, 146, 149, 155, 163, 166, 170, 179, 184, 232, 271-274
core metadata 90, 92, 93, 187
cost function 122, 123
coverage 54, 55, 67, 76, 96, 97, 99, 101, 102, 110, 111, 115, 125-127, 132, **165-177,** 179-181, 209, 213, 214
CubeWerx 213, 214
cubic spline (CubicSpline) 138, 139, 178
cultural adaptability 22, 23, 24
curve (Curve, _Curve) 35, 37, 75, 76, 82, 115, 126, 128, 130-145,

Index 311

147-149, 153, 154, 157-162, 166, 167, 169, 171, 172, 174-180, 184, 232, 233
CV_CommonPointRule 176
CV_ContinuousCoverage 165, 174
CV_ContinuousQuadrilateralGrid Coverage 165, 170, 174
CV_Coverage 165, 174
CV_DiscreteCoverage 165, 174
CV_DiscreteCurveCoverage 165, 166, 174
CV_DiscreteGridPointCoverage 165, 168, 174
CV_DiscretePointCoverage 165, 166, 168, 174
CV_DiscreteSolidCoverage 165, 167, 174
CV_DiscreteSurfaceCoverage 165, 167, 174
CV_DomainObject 172
CV_GeometryValuePair 165, 172
CV_Grid 165
CV_GridPointValuePair 165, 127, 172
CV_GridValuesMatrix 127, 165, 168, 173
CV_HexagonalGridCoverage 165, 170, 174
CV_PointValuePair 127, 165
CV_SegmentedCurveCoverage 165, 171, 174, 175
CV_ThiessenPolygonCoverage 165, 169, 174
CV_TINCoverage 165, 171, 174, 175
CV_ValueObject 165, 172
cylinder 132, 142

D

data dictionary 54, 114
data exchange 22, 27, 41, 87, 88, 188, 194, 195, 215
Data management and interchange (JTC1/SC32) 186, 191
data model 38, 44-46, 83, 84, 96, 99, 102, 125, 195, 204
data product specification 54, 115
data quality measure 90, 95, 96, 234
data type 39, 56-58, 87, 114, 182, 215
datum 39, 77-81, 235
de-facto standard 4, 6, 8, 9, 16, 49, 51, 96, 97, 101
de-jure standard 4
deliverables 14, 19, 30, 32
Deutsches Institut für Normung 4, 10, 214, 216, 286, 299, 301
DGIWG 30, 34, 51, 71, 101, 114, 193-195, 301, 302
dictionary 54, 58, 94, 114, 177, 183
difference 152
DIGEST 49, 71, 101, 194, 301
Digital Geographic Information Working Group 30, 34, 51, 71, 101, 114, 193-195, 301, 302
DIN 4, 10, 214, 216, 286, 299, 301
direct position (DirectPosition) 126, 127, 146, 149, 151, 153, 156, 176, 236
DIS 13-15, 32, 50, 212, 225, 301
disaster management 27, 52, 209
discrete coverage 166-168, 174, 176, 181
distance 137, **148,** 150, 154, 157, 177
distribution 42, 71, 89, 90, 96, 215
Document description and processing languages (JTC1/SC34) 186, 191
Document Object Model (DOM) 299, 301
Document processing and related communication (JTC1/SC18) 186, 190
Document Type Declaration (DTD) 88, 89, 287, **290-292,** 301
DOM 299, 301
domain object 172, 177
domain reference model 45, 46, 72
Draft International Standard (DIS) 13-15, 32, 50, 212, 225, 301
DTD 88, 89, 287, **290-292,** 301
Dublin core 91-93, 184, 187

E

Earth Gravity Model (EGM) 82
Ecosystem Associates 209
ellipse 106, 119, 132, 136
ellipsoid 77-81, 134
encoding 7, 32, 41, 49, 50, 71, **87-89**, 97, 102, 184, 187, 213-216, 238, 286, 296
engineering coordinate system 39, 79
engineering datum 79, 238
engineering viewpoint 42, 43
enquiry stage 13, 14, 18, 50
enterprise viewpoint 42, 43, 190
enumeration 58, 60, 63, 170
envelope 146, 149
Environment Directorate General 30
environment 27, 29, 52, 71, 200, 201, 203, 204, 209
EOSE 47, 72, 75, 301
EPSG 193, 195, 301
equals 151
ESRI 207, 209, 301
Europe 1, 4, 6, 27-30, 51, 81, 192-196, 201, 202, 205, 211
European Petroleum Survey Group (EPSG) 193, 195, 301
evaluateInverse 174, 175
evaluation 175, 176
exchange format 27, 41
EXPRESS 56, 188
Extended Open System Environment Reference Model (EOSE-RM) 72, 75
Extended Open Systems Environment (EOSE) 47, 72, 75, 301
Extensible Markup Language (XML) 5, 6, 41 50, 87-89, 108, 177, 182, 184, 215, 216, **286-299**, 306
Extensible Stylesheet Language (XSL) 299, 306
external function 86, 239
external liaison member 31, 32, **192-206**, 211

F

FACC data dictionary 114, 301
FAO 193, 195, 196
FDIS 13, 14, 32, 50, 212, 301
feature 32, 37-41, 44, 45, 61, 72, 75-77, 79, 82, 85-89, 94, 95, 99, 110-116, 119, 125, 131, 159, 163, 164, 166, 172, 175-177, 179, 181, 183, 209, 215, 216, 239, 240
feature catalogue 38-40, 44, 54, 75, 76, 86, 89, 112-114, 213, 215, 240
Federal Geographic Data Committee (FGDC) 91, 92, 196, 207, 209, 213, 301
Fédération Internationale des Géomètre 124, 192, 193, 199, 301
FGDC 91, 92, 196, 207, 209, 213, 301
FIG 124, 192, 193, 199, 301
Final Draft International Standard (FDIS) 13, 14, 32, 50, 212, 301
fleet management 27, 188
Flexible magnetic media for digital data interchange (JTC1/SC11) 186, 190
frame camera 102, 105, 110
framework 8, 31, 49, 51, 72, 96, 108, 125, 190, 207
functional standard 47, 48, 53, **71**, 195, 241

G

Galdos Systems 209, 213, 214
Gauss-Krüger System 80, 81
gazetteer 82, 241, 242
GDF 30, 71, 189, 301
General Dynamics 207
general feature model 37, 44, 75
geodesic 133, 134, 154, 155, 178
geodetic codes and parameters 40, 54, 77, **81**
geographic coordinate 77
Geographic Data Files (GDF) 30, 71, 189, 301

Index 313

geographic feature 75, 77, 82, 166, 179, 183, 242
geographic identifier 30, 54, 77, **82,** 242
Geographic Information Human Interaction Service (GIHS) 73, 75
Geographic Information Model Management Service (GIMS) 73, 75
Geographic Information Processing Service (GIPS) 73, 75
Geographic information service 45-47, 61, 72, 75
Geographic Resources Analysis Support System (GRASS) 207, 302
Geography Markup Language (GML) 33, 48, 49, 67, 77, 87, 125, **177-184**, 209-211, 213-214, 216, 302
geometry (Geometry) 159, 163, 179, 182
geometry value pair 172, 177
geometry collection (GeometryCollection) 159, 161
georeference 54, 71, **77,** 99, 100, 101, 104, 106, 108
Geospatial Fusion Services Pilot Project 211
GeoTIFF 5, 97, 101, 195
GetCapabilities 116
GetFeatureInfo 116
GetMap 116
GF_FeatureType 75, 166
GIF 101
GKS 27-29, 190, 191
Global Navigation Satellite System (GNSS) 83, 106, 189, 302
Global Spatial Data Infrastructure (GSDI) 51, 52, 193, 196, 197, 302
GM_AffinePlacement 137
GM_Aggregate 129-131, 159
GM_Arc 129, 135
GM_ArcString 129, 134, 135
GM_ArcStringByBulge 129, 135, 136
GM_Bezier 129, 140
GM_BicubicGrid 129, 132, 144

GM_BilinearGrid 129, 132, 143
GM_BSpline 139
GM_BSplineSurface 129, 132, 144, 145
GM_Circle 129, 135
GM_Clothoid 129, 137
GM_Complex 129, 130
GM_CompositeCurve 129, 130
GM_CompositeSurface 129, 130
GM_Cone 129, 132, 142
GM_Conic 129, 136
GM_CubicSpline 129, 138, 139
GM_Curve 127, 129-131
GM_CurveSegment 129, 131
GM_Cylinder 129, 132, 142
GM_Geodesic 129, 134
GM_GeodesicString 129, 134
GM_GriddedSurface 129, 132, 141
GM_LineSegment 129, 133, 154
GM_LineString 129, 133, 147
GM_Object 126, 129, 146, 148-153, 159, 172
GM_OffsetCurve 129, 137, 184
GM_OrientableCurve 129
GM_OrientableSurface 129, 131
GM_ParametricCurveSurface 131, 141
GM_Placement 137
GM_Point 39, 126, 127, 129, 130, 153, 154, 165
GM_PointArray 126, 127, 135, 154
GM_PointObject 127
GM_Polygon 129, 131, 140, 145
GM_PolyhedralSurface 129, 145
GM_PolynomialSpline 129, 138
GM_Position 126, 127, 153
GM_Primitive 129, 153
GM_Solid 127, 129, 130, 153
GM_Sphere 129, 132, 143
GM_SplineCurve 129, 137
GM_Surface 126, 127, 129-131, 145
GM_SurfacePatch 129, 131
GM_Tin 129, 131, 132, 145, 146, 171
GM_Triangle 129, 140, 141, 145

GM_TriangulatedSurface   129, 132, 145
GML   33, 48, 49, 67, 77, 87, 125, **177-184,** 209-211, 213-216, 302
Graphical Kernel System   27-29, 190, 191
Graphics standard   28
GRASS   207, 302
grid values matrix   127, 168
GRSS   193, 197, 302
GSDI   51, 52, 193, 196, 197, 302

H
Harmonized Model Maintenance Group (HMMG)   32, 33
HDF   101, 302
hexagonal grid   97, 111, 170, 171
Hilbert order   174
HMMG   32, 33
homeland security   52
horizontalCurve   157
HTML   29, 191, 287, 302
Human Interaction Service (ITHS)   73
Human Technology Interface (HTI)   73, 74
hydrographic sonar   102-104
Hypertext Markup Language   29, 191, 287, 302

I
IAG   192, 193, 197, 198, 302
IBM   56, 302
ICA   10, 193, 198, 302
ICAO   193, 198, 199, 302
IEC   9, 18, 19, 25, 28, 31, 42, 185-187, 189-192, 302
IEEE Geoscience and Remote Sensing Society   193, 197
IHB   193, 302
IHO   30, 51, 71, 192, 193, 200, 302
imager   109
imagery   5, 16, 32, 34, 37, 49, 53, 54, 90, 94, **96-111,** 125, 201, 207
implementation   5, 7, 21, 27, 31, 33, 42, 48-50, 52, 54, 56-58, 67, 68, 89, 108, **115,** 177, 183, 184, 196, 201, 207, 208, 210, 211, 214, 215, 231, 239, **245**
Implementation Specification   31, 50, 183, 207, 208, 210, 211, 215
implementation under test (IUT)   67, 68, 303
Industrial automation systems and integration   56, 186, 188
inertial system   83, 245, 246
Information Service Interface (ISI)   73, 74
Information Technology   1, 4-7, 17, 18, 28, 29, 31, 41-43, 47, 67, 70-75, 88, 92, 186, 191, 209
information viewpoint   42, 43, 177
Infrastructure for spatial information in Europe (INSPIRE)   30, 34, 51, 52, 202, 302
infrastructure standard   55, 71
INSPIRE   30, 34, 51, 52, 202, 302
Instance   82, 89, 126, 246
Instance model   89, **246**
Instant   77, 109, 180, **246**
Interconnection of information technology equipment (JTC1/SC25)   186, 191
Intergraph   207
Interlis   49
internal liaison member   11, 31-33, 185-192
International Association of Geodesy (IAG)   192, 193, 197, 198, 302
International Cartographic Association (ICA)   10, 193, 198, 302
International Civil Aviation Organisation (ICAO)   193, 198, 199, 302
International Electrotechnical Organization (IEC)   9, 18, 19, 25, 28, 31, 42, 185-187, 189-192, 302
International Federation of Surveyors (FIG)   124, 192, 193, 199, 301

Index 315

International Federation of the National Standardising Associations (ISA) 9, 302
International Hydrographic Bureau (IHB) 193, 302
International Hydrographic Organization (IHO) 30, 51, 71, 192, 193, 200, 302
International Interfaces 209
International Society for Photogrammetry and Remote Sensing (ISPRS) 5, 10, 11, 193, 194, 200, 201, 302
International Standard (IS) 6, 7, 10, 13-16, 18, 25, 27, 32, 34, 50, 51, 56, 68-71, 111, 120, 184, 200, 211, 212, 214, 302
International Standardised Profile (ISP) 69, 70
International Steering Committee for Global Mapping (ISCGM) 193, 196, 201, 302
International Telecommunication Union (ITU) 9, 18, 19, 25, 185, 186, 192, 198, 202, 208, 303
International Terrestrial Reference System (ITRS) 81, 82, 303
International Workshop Agreement (IWA) 14, 16, 303
interoperability 4, 22, 50, 177, 194, 202, 208, 210, 211, 215, 216, 246, 247, 287
Interoperability Program 210, 211
intersection 127, 151, 152
IS 6, 7, 10, 13-16, 18, 25, 27, 32, 34, 50, 51, 56, 68-71, 111, 120, 184, 200, 211, 212, 214, 302
ISA 9, 302
ISCGM 193, 196, 201, 302
isCycle 148
ISO 6709 77
ISO 10011 20
ISO 11180 120
ISO 19011 20
ISO 19101 36, **41-49**, 53, 55, 68, 72, 96

ISO 19102 55
ISO 19104 53, 55, **66,** 72, 102
ISO 19105 16, 32, 53, 55, **67, 68**
ISO 19106 53, 55, **68-70,** 114
ISO 19107 39, 48, 49, 53, 55, 60, 64, 65, 68-70, 72, 73, 76, 97, 115, **125-159,** 163, 164, 177, 183, 184, 215
ISO 19108 49, 53, 70, **76,** 180, 184
ISO 19109 37, 53, 68, 73, **75,** 76, 113, 166, 184, 214
ISO 19110 38, 49, 54, 73, 76, **112-114,** 213-215, 275
ISO 19111 39, 49, 54, 63, 64, 70, 73, **77-84,** 97, 120, 133, 166, 179, 184, 215, 272
ISO 19112 54, 73, 77, **82,** 272
ISO 19113 36, 54, 90, **93,** 94, 96, 97, 102
ISO 19114 36, 37, 54, 90, 93-97
ISO 19115 12, 24, 36, 37, 45, 49, 50, 54, 70, 73, **89-94,** 102, 184, 213-216, **260-286,** 308, 309
ISO 19116 54, 66, 73, 77, **83, 84**
ISO 19117 40, 53, 73, **85,** 97, 184
ISO 19118 41, 49, 53, **87-89,** 97, 184, 213-216, 286
ISO 19119 53, 72, 83, 213-214
ISO 19123 48, 54, 55, 67, 73, 76, 96, 97, 101, 102, 110, 111, 115, 125-128, 132, **165-177,** 180, 181, 213, 214
ISO 19125-1 54, 55, 73, 115, 125, 126, **159-164**
ISO 19126 54, 114
ISO 19127 40, 54, 77, **81**
ISO 19128 54, 73, 116, 211, 213, 214
ISO 19129 54, 96-101, 125
ISO 19130 54, 96, 99, 101, **102-111,** 125
ISO 19131 54, 115
ISO 19132 54, 117
ISO 19133 48, 54, **117-123**
ISO 19134 54, 117, 118

ISO 19135   23, 40, 51, 54, 81, 95, **111,** 112, 114
ISO 19136   49, 54, 55, 67, 77, 87, 125, 159, **177-184,** 211, 213, 216
ISO 19137   49
ISO 19139   50, 54, 184, 211
ISO 19140   50
ISO 9000   1, 3, 8, 19-22
ISO 9001   20, 21
ISO 9004   20, 22
ISO/IEC 10746   42
ISO/IEC 11179   92
ISO/IEC 14481   43, 45
ISO/IEC DIS 19501   56
ISO/IEC JTC1   18, 19, 28, 29, 56, 69, 77, 101, 186, **190-192,** 303
ISO/IEC TR 14252   47, 72
ISO/RS 19124   54, 96, 101, 102
ISO/TC12   186, 187
ISO/TC20   186, 187
ISO/TC23   186, 187
ISO/TC37   186, 187
ISO/TC42   186, 187
ISO/TC46   186-188
ISO/TC82   185, 186, 188
ISO/TC130   186, 188
ISO/TC172   186, 188
ISO/TC176   20, 21, 186, 188
ISO/TC184   56, 186, 188
ISO/TC204   31, 33, 71, 118, 185, 186, 188, 189
ISO/TC211   1, 6, 11, 12, 22, 23, 27, 29-34, 48-52, 67, 71, 77, 101, 102, 123, 179, 184-189, 192, 210, 211, 213-216
ISO/TR 19120   47, 48, 53, **71,** 195, 241
ISO/TR 19121   16, 32, 54, 96, **101**
ISO/TR 19122   54, 123, 124
ISO/TS 19103   15, 53, **55-60,** 214
ISO/TS 19138   90, 93, 95, 96
ISPRS   5, 10, 11, 193, 194, 200, 201, 302
isSimple   148

IT   1, 4-7, 17, 18, 28, 29, 31, 41-43, 47, 67, 70-75, 88, 92, 186, 191, 209
IT security techniques (JTC1/SC27)   186, 191
ITU   9, 18, 19, 25, 185, 186, 192, 198, 202, 208, 303
IWA   14, 16, 303

J
Jacobson   56
Java   15
Joint Research Centre of the European Union (JRC)   30, 192, 193, 201, 202, 303
Joint Technical Committee 1   18, 19, 28, 29, 56, 69, 77, 101, 186, 190-192, 303
JPEG   96, 101, 191, 303
JRC   30, 192, 193, 201, 202, 303
JTC1   18, 19, 28, 29, 56, 69, 77, 101, 186, **190-192**, 303
JTC1/SC2   186, 190
JTC1/SC7   186, 190
JTC1/SC11   186, 190
JTC1/SC18   186, 190
JTC1/SC20   12, 186, 187
JTC1/SC22   186, 190
JTC1/SC23   186, 190
JTC1/SC24   29, 186, 190, 191
JTC1/SC25   186, 191
JTC1/SC27   186, 191
JTC1/SC29   186, 191
JTC1/SC32   29, 186, 191
JTC1/SC34   29, 186, 191
JTC1/SC35   186, 191, 192

K
Keplerian ellipse   106
knot (Knot)   128, 140, 178

L
latitude   77, 81, 107, 143
LBS   33, 34, 48-50, 117-123, 303
liaison member   1, 11, 31, 32, 51, 101, 185-212

lidar 102, 103, 109
line segment 133, 154
line string (LineString) 133, 154, 159, 160, 178
lineage 37, 94, 265
linear reference system 119, 120, 248
linear scanning 173
line-of-sight sensor 109
line-scanners 109
linguistic adaptability 1, 6, 22, 23, 24
linking element 293
locate 174, 175
location based services 33, 34, 48-50, 117-123, 303
location instance 82
location type 82
Lockheed Martin 207, 209
logical consistency 37, 94
longitude 77, 78, 80, 81, 107, 143

M

Management Committee 208, 210, 303
map projection 77, 79, 132, 133, 248
Maritime navigation and radio-communication equipment and systems (IEC/TC80) 186, 189
mbRegion 146
meetings 308
meta model 43-46
metadata 7, 12, 23, 30, 32, 36, 37, 45, 46, 49, 50, 52, 61, 71, 76, 85, **89-94,** 102, 116, 119, 184, 187, 190, 209, 211, 213-216, 248, 249, 260-286
metadata – implementation specification 54
meta-meta model 43, 44
methodology for feature cataloguing 38, 112-114, 213-215
Microsoft 3, 101, 303
Mining 185, 186, 188
Model Management Service (ITMS) 73
Morton order 174

MPEG-2 101
multi curve (MultiCurve) 159-162, 178
multimodal location based services 117, 118
multi point (MultiPoint) 159, 160, 178, 181
multi surface (MultiSurface) 159, 163, 181

N

namespace 182, 183
NASA 101, 194, 207, 209, 303, 305
National Aeronautics and Space Administration (NASA) 101, 194, 207, 209, 303, 305
National Imagery and Mapping Agency (NIMA) 101, 207, 304
National Information Standards Organization (NISO) 304
national member 31, 51, 198
NATO 30, 49, 51, 71, 101, 114, 194, 301, 303, 304
navigation 48, 71, 78, 83, 86, 106, 118-123, 134, 186, 189, 198, 199, 200
Network-to-Network interface 74
New Work Item Proposal (NWIP) 13, 15, 49, 96, 212, 304
NIMA 101, 207, 304
NITF 101, 304
node 77, 118, 120-122, 158, 159, 180, 182, 249
non-rigorous model 110
North Rhine Westfalia Pilot Project (NRWPP) 211
Northrop Grumman 207
NWIP 13, 15, 49, 96, 212, 304

O

Object Constraint Language (OCL) 57, 58, 304
observing member (O-member) 31, 32
OCL 57, 58, 304

offset 119, 120, 131, 133, 135, 137, 170
OGC 1, 5, 8, 11, 31, 33, 48, 49, 51, 67, 110, 115, 116, 179, 180, 183, 184, 193, 197, 202, **207-212,** 213, 215, 216, 303, 304
O-member 31, 32
OMG 56, 304
Open Distributed Processing Reference model 42, 43, 190
Open GIS Consortium 1, 5, 8, 11, 31, 33, 48, 49, 51, 67, 110, 115, 116, 179, 180, 183, 184, 193, 197, 202, **207-212,** 213, 215, 216, 303, 304
Open System Environment Reference Model (OSE-RM) 47, 72, 73, 304
operation 37-39, 43, 44, 57-59, 61, 75, 79, 83, 86, 103, 111, 113, 116, 146-157, 174-177, 215, 216
Optical disk cartridge for information interchange (JTC1/SC23) 186, 190
Optics and optical instruments 186, 188
Oracle 209
orthophoto 99
Østensen, Olaf 31, 34
outreach 31, 33, 51, 52
overview 55

P

package-diagram 56, 59, 64, 65, 91
paper and film scanners 105
participating member (P-member) 13-15, 31, 32
PAS 14, 15, 304
PC IDEA 193, 197, 203, 204, 304
PCGIAP 192, 193, 197, 203, 304
PCI Geomatics 209, 213, 214
PDA 118, 304
perimeter 157
period 251, 180
Permanent Committee on GIS Infrastructure for Asia and the Pacific (PCGIAP) 192, 193, 197, 203, 304
Permanent Committee on Spatial Data Infrastructure for Americas (PC IDEA) 193, 197, 203, 204, 304
Personal Digital Assistant (PDA) 118, 304
PHIGS 28, 29, 191, 304
PhotoCampactDisk 101
placement 137
platform 15, 22, 41, 87, 98, 103, 104, **106, 109, 110,** 115, 202, 210, 215
P-member 13-15, 31, 32
PNG 29, 101, 191, 304
point (Point) 35, 37, 70, 75, 82, 115, 126-128, 130-132, 134, 135, 138-142, 145-148, 150, 151, 153-160, 162-164 166, 168, 169, 172, 173, 178, 179, 180, 184, 251
polygon (Polygon) 131, 132, 140, 145, 146, 159, 162, 163, 169, 175, 178
polyhedral surface 132, 145
polynomial spline (polynomialSpline) 132, 137, 138, 144, 178
Portable Network Graphics 29, 101, 191, 304
portrayal 32, 40, 52, 71, 76, **85, 86,** 97, 163, 184, 251, 252
portrayal catalogue 40, 85, 86, 251, 276, 277
positional accuracy 37, 93, 94, 252
positioning service 54, 66, 77, **83, 84**
Postscript 28
Preparatory stage 13, 15, 18
prime meridian 39, 78, 252
Processing Service (ITPS) 73, 74
profile 6, 30, 31, 45, 47, 50, 53, 55, **68-70,** 111, 114, 138, 182, 183, 215, 252, 302
profiler 109

Programmer's Hierarchical Interactive Graphic System 28, 29, 191, 304
Programming languages, their environment and systems software interfaces (JTC1/SC22) 186, 190
Project Team (PT) 12
projected coordinate system 77-81, 116
Proposal stage 13, 18
Publication stage 14, 18
Publicly Available Specification (PAS) 14, 15, 304

Q
quadrilateral grid 97, 111, 127, 132, 169, 170, 174, 175
qualification of personnel 54, 123, 124, 199
quality 8, 19-22, 28, 30, 32, 36, 37, 44, 52, 62, 71, 75, 84, 89, 90, **93-96**, 119, 186, 188, 214, 215, 233-235, 252, 253, 264-268
quality evaluation procedure 36, 37, 90, 93-97
quality evaluation report 36, 94, 95
quality management and quality assurance 20, 21, 186, 188
quality management system 3, 19-22
quality principle 36, 90, 93, 94, 96, 97, 102
Quantities, units, symbols, conversion factors 186, 187

R
radius 135, 137, 143, 155
range 167, 168, 176, 184
rapid positioning coordinate 109, 110
raster 28, 90, 96, 101, 125, 194
Rational Software Corporation 56
Reference model 36, **41-49**, 53, 55, 68, 72, 96
Reference Model of Open Distributed Processing (RM-ODP) 43, 72
refinement 59, 209

registration 21, 23, 40, 54, 102, **111, 112**, 114
registry 40, 51, 81, 95, **111, 112**
remote sensing 6, 11, 99, 102, 108, 193, 194, 197, 200, 201, 213
representativePoint 146, 147
Request for Comment (RFC) 208, 209, 304
Request for Information (RFI) 208, 209, 304
Request for Proposal (RFP) 208, 209, 212, 304
Request for Technology (RFT) 208, 304
Review Summary 32, 96, 101
RFC 208, 209, 304
RFI 208, 209, 304
RFP 208, 209, 212, 304
RFT 208, 209, 304
Riemann 97, 98
rigorous sensor model 106, 108, 109
route request 122
routing 117, 118
RS 32, 96, 101
rules for application schema 37, 75, 76, 184
Rumbaugh 56

S
S-57 71, 101, 305
Sæterøy, Bjørnhild 11
samplePoint 154
SC 8, 11-13, 15, 17, 28, 29, 31, 32, 185, 187, 188, 305
Scalable Vector Graphics (SVG) 28, 299, 305
scanning linear array 102, 104
SCAR 193, 204, 205, 305
SCC 10, 305
Schema for coverage geometry and functions 48, 54, 55, 67, 73, 76, 96, 97, 101, 102, 110, 111, 115, 125-128, 132, **165-177**, 180, 181, 213, 214

Scientific Committee on Antarctic Research (SCAR) 193, 204, 205, 305
SDI 30, 51, 52, 193, 196, 197, 203, 204, 302, 304, 305
segmented curve coverage 111, 169, 171, 172, 176
self-description principle 45
semantic interoperability 22
sensor 97, 99, 100, **102-105,** 201
sensor model 96, 102-108, 125
SensorML 5, 108-110
service 8, 19, 21, 31, 34, 42, 43, 45, 47, 48, 50, 53, 54, 61, 66, **71-75,** 77, 83, 84, 87, 88, 102, 109, 116-119, 122, 177, 180, 186, 192-194, 199, 202, 206, 209-211, 213, 214, 238, 242, 251, 254
service standard 72
Set Theory 128, 151-153
SGML 29, 190, 191, 287, 297, 305
SICAD Geomatics 209
SIG 8, 209, 305
simple feature 54, 55 115, 125, **159-164,** 209
SLR camera 110
Software and systems engineering (JTC1/SC7) 186, 190
solid (Solid) 126, 130, 132, 158, 167, 178, 180, 184
Spatial Data Infrastructure 30, 51, 52, 193, 196, 197, 203, 204, 302, 304, 305
spatial referencing 39, 49, 63, 64, 70, 77-84, 97, 120, 166, 179, 184, 215
spatial schema 32, 55, 60, 64, 65, 70, 76, 97, 115, **125-159,** 163, 164, 177, 183, 184, 215
spatiotemporal domain 166, 167, 170, 172, 175
Special Interest Group (SIG) 8, 209, 305
sphere 131-133, 143, 148
spiral scanning 173
spline 126, 128, 131-133, 137-140, 144, 145
SQL 85, 115, 191, 305
SQL/MM 29, 101, 191, 305
SRI 209
Standard Generalised Markup Language 29, 190, 191, 287, 297, 305
Standardisation Council of Canada (SCC) 10, 305
Standards Developing Organizations (SDO) 1, 8, 9, 18, 19, 25, 305
stereotype 56, 59, 60, 255, 256
strategic member 207, 209, 210
strategy 6, 31, 33, 51, 176
Structured Query Language (SQL) 85, 115, 191, 305
Structured Query Language/MultiMedia 29, 101, 191, 305
Subcommittee (SC) 8, 11-13, 15, 17, 28, 29, 31, 32, 185, 187, 188, 305
Sun Microsystems 15
surface (Surface, _Surface) 35, 37, 75, 82, 115, 126, 128, 130-132, 134, 140-146, 149, 156, 157, 159, 162, 167, 178, 184, 215, 256
SVG 28, 299, 305
symbolisation 85, 86
symmetricDifference 153
Synthetic Aperture Radar (SAR) 102, 103, 305
System Management Service (ITSS) 73

T

tangent 138, 154
Task Force ISO/TC211 – ISO/TC204 33
TC 8, 11-13, 16, 17, 20, 21, 29, 31, 32, 185, 192, 208-210, 305
Technical Committee (TC) 8, 11-13, 16, 17, 20, 21, 29, 31, 32, 185, 192, 208-210, 305
Technical Report 14, 15, 32, 71, 96, 101, 123, 306

Technical Specification (TS) 14, 15, 81, 95, 306
technology viewpoint 42, 43
temporal accuracy 37, 94
temporal schema 32, 70, 76, 89, 180, 184
terminology 8, 17, 25, 36, 37, 43, 53, **66, 67,** 72, 102, 55, 125-127, 186, 187, 189
Terminology Maintenance Group (TMG) 33
tessellation 98, 146, 170
thematic accuracy 37, 94
Thiessen polygon 97, 111, 146, 169, 175
TIFF 3, 5, 67, 96, 97, 101, 188, 305
TIFF/IT 101, 188, 305
TIN (tin) 97, 111, 131, 132, 146, 169, 171, 172, 174, 175, 178, 305
TMG 33
topology 39, 158, 159, 163, 164, 179, 180, 182, 215
Total Station 83
tourism GIS 35-41, 44
TP_Object 158
TR 14, 15, 32, 71, 96, 101, 123, 306
tracking 48, 118-120
Tractors and machinery for agriculture and forestry 186, 187
transformation 5, 22, 58, 77, 79-81, 84, 97, 98, 104, 117, 133, 137, 149, 170, 195, 209
travelling salesman 118
Triangulated Irregular Network (TIN, tin) 97, 111, 131, 132, 146, 169, 171, 172, 174, 175, 178, 305
triangulated surface 132, 145, 146
trigger 119
TS 14, 15, 81, 95, 306

U

U.S. Army Corps of Engineers 209
UCGIS 124
UML 38, 43, 44, **55-67,** 73, 88, 91, 126, 129, 214, 215, 230, 255, 256, 298, 306

UN 18, 19, 51, 192-196, 199, 203-206, 306
UNECE 193, 205, 306
UNGEGN 192, 193, 206, 306
UNGIWG 193, 205, 306
Unified Modelling language (UML) 38, 43, 44, **55-67,** 73, 88, 91, 126, 129, 214, 215, 230, 255, 256, 298, 306
Uniform Resource Locator 116, 306
union 152
unit of measurement 58, 259
United Nations 18, 19, 51, 192-196, 199, 203-206, 306
United Nations Economical Commission for Europe, Statistical Division (UNECE) 193, 205, 306
United Nations Geographic Information Working Group (UNGIWG) 193, 205, 306
United Nations Group of Experts on Geographical Names (UNGEGN) 192, 193, 206, 306
Universal Transverse Mercator (UTM) 78-82, 306
University Consortium for Geographic Information Science 124
upNormal 156
URL 116, 306
usage 37, 59, 70, 94
use-case diagram 66
User interfaces (JTC1/SC35) 186, 191, 192
USGS 209
UTM 78-82, 306

V

value object 172, 175
verticalCurve 132, 143, 144, 157
video camera 110
viewpoint 6, 42, 43, 72, 177
Virtual Reality Modelling Language 29, 49, 306
VRML 29, 49, 306

W

W3C   9, 181, 306
WD   13, 210, 306
Web Feature Server   215, 306
Web Map server   54, 115-117, 211, 213, 214, 306
Web Mapping Testbed   211
WG   5, 8, 12, 13, 15, 25, 28, 30-34, 185, 187-191, 193, 196, 208, 209, 306
WGS84   78, 80
WMO   192, 193, 206, 306
Workflow/Task Service (ITWS)   73
Working Draft (WD)   13, 210, 306
Working Group   5, 8, 12, 13, 15, 25, 28, 30-34, 185, 187-191, 193, 196, 208, 209, 306
World Meteorological Organisation (WMO)   192, 193, 206, 306
World Wide Web consortium (W3C)   9, 181, 306

X

XMI   298, 306
XML   5, 6, 41 50, 87-89, 91, 108, 177, 182-184, 215, 216, **286-199,** 306
XML schema   88, 177, 183, 215, 288, 289
XSL   299, 306

Be the first to know
with the new online notification service

Springer Alert

You decide how we keep you up to date on new publications:
- Select a specialist field within a subject area
- Take your pick from various information formats
- Choose how often you'd like to be informed

And receive customised information to suit your needs

http://www.springer.de/alert

---

and then you are one click away from a world of geoscience information!

Come and visit Springer's **Geoscience** Online Library

http://www.springer.de/geo

Made in the USA
Lexington, KY
26 August 2011